The DEVIL — AND THE — DEEP BLUE SEA

Also by Robert Holden

D. H. Souter's Cats

The Mechanical Eye and the Illustrated Book

A Golden Age: Australia's Fantasy Illustrators [2 volumes]

Twinkle, Twinkle Southern Cross: The Forgotten Folklore of Australian Nursery Rhymes

Cover Up: The Art of Magazine Covers in Australia

Orphans of History: The forgotten children of the First Fleet

Contributing Author

Bunyips: Australia's Folklore of Fear, with Nicholas Holden

Hall Thorpe Colour Woodcuts, with Ingrid Holden

The Fairy World of Ida Rentoul Outhwaite, with Marcie Muir

The Art of Frederick McCubbin [introductory essay]

'Banjo' Paterson's *Song of the Wheat* [introductory essay]

The Artist and the Patron

The Legendary Lindsays

Doll's Houses in Australia, 1870 - 1950

The World Upside Down: Australia 1788 - 1830

The Endless Playground: Celebrating Australian Childhood

For Children

A Treasury of Children's Verse

Verse Ahoy, illustrated by Heather Potter

The Pied Piper of Hamelin, illustrated by Drahos Zak

The DEVIL — AND THE — DEEP BLUE SEA

ROBERT HOLDEN

The life of the Australian whaling captain, William Chamberlain: A tale of abduction, adventure and murder

HarperCollins*Publishers*

HarperCollins*Publishers*

First published in Australia in 2003
by HarperCollinsPublishers Pty Limited
ABN 36 009 913 517
A member of the HarperCollinsPublishers (Australia) Pty Limited Group
www.harpercollins.com.au

HarperCollinsPublishers
25 Ryde Road, Pymble, Sydney, NSW 2073, Australia
31 View Road, Glenfield, Auckland 10, New Zealand
77–85 Fulham Palace Road, London W6 8JB, United Kingdom
2 Bloor Street East, 20th floor, Toronto, Ontario M4W 1A8, Canada
10 East 53rd Street, New York NY 10022, USA

National Library of Australia Cataloguing-in-Publication data:

Holden, Robert.
The devil and the deep blue sea: the life of Australian whaling captain, William Chamberlain: a tale of abduction, adventure and murder.
Bibliography.
ISBN 0 7322 7557 1.
1. Chamberlain, William. 2. Whaling masters – Australia – Biography. 3. Whaling – Australia – History. I. Title.
639.28092

Cover and internal design by Gayna Murphy, HarperCollins Design Studio
Cover illustration: map from *A Complete Account of the Settlement at Port Jackson*, Watkin Tench, 1793; whale flukes from *Journal of a Voyage to Greenland in the Year 1821*, G. W. Manby, 1824; sea and boat from a facsimile edition of *Two Years in the Pacific & Arctic Ocean*, James F. Munger, 1852
Typeset in 12 on 16 Historical-Fell Type by HarperCollins Design Studio
Printed and bound in Australia by Griffin Press on 79gsm Bulky White

To Rosalind and Ian Hollinrake,
for their enduring friendship

Contents

PART THREE

AUTHOR'S NOTE

The Lord with his sore, and great, and strong sword,
shall punish Leviathan ... He shall slay the dragon
that is in the sea.

ISAIAH, OLD TESTAMENT

The initial inspiration for *The Devil and the Deep Blue Sea* overwhelmed me on a fateful day in 2000. One moment I was completing research for a book to be published by the National Library of Australia on the bunyip. The next moment I was transfixed by a mere half-page of text; and this was something totally unrelated to the work in progress.

Here I read the tantalisingly brief interview that William Chamberlain gave to the Navy Board in Portsmouth in late 1816. The moving words of this twelve year old continued to reverberate in my thoughts for days thereafter:

'How long is it since you left Port Jackson?'
'I believe it is rather more than four years.'
'Did your father or mother know of your going away?'
'No, Sir."
How came you to leave ...?'

'I was taken away by the Captain of the Frederick . . . he made me very ill, until we were taken by the Privateer.'

My own reaction echoed that of the boy as I too was 'taken away' by his story. Here was abduction, pitiable youth, grieving parents, pirates and . . . a mystery. Why had this boy been taken and what role had he played in his four years away that made him an object of interest and concern to the Navy Board?

These were questions not easily answered but their insistent clamour would not be stilled. I began to follow a research trail that would unravel this mystery. Yet even as a trained librarian and experienced researcher this was a daunting task. However, when I was faced with continual blanks in the historical narrative, this only became a further incentive urging me on.

My imagination kept replaying the reunion [more a resurrection, really] that occurred in Sydney-town one day early in 1817, when the boy returned home after five years as if back from the dead. And returned, moreover, as a young naval hero bearing letters to the Governor of the Colony. Eventually I was able to place William in the celebrated context of being the third and youngest Australian to join the Royal Navy. [His immediate predecessors were Robert Johnston, the second son of celebrated First Fleeters, and Phillip Parker King. These two boys joined the Navy in 1807 but both were at least three years older than William was at the time he joined.]

Then I unearthed a revelation — a manuscript in the private papers of Governor Macquarie dated November 1810. Hidden away in this voluminous collection in the Mitchell Library was a short entry which gave the whole saga a new dimension. There I found that the whaling captain who had abducted young William had been declared 'labouring under serious mental derangement' by the Colony's acting principal surgeon! Within a year, though, this deranged man was let loose for his return voyage to England with a valuable cargo of whale oil. And it was then that he abducted William, not quite eight years old.

Parallels with the insane Captain Ahab from *Moby Dick* flooded my mind. If I had needed any further incentive to pursue this story, I had found it here. But in this first excited flush of research as I discovered an almost unbelievable drama, I wondered if anyone would credit that it was a real-life story. After all, there are times when it reads like a 'Boy's Own' adventure!

These unparalleled exploits brought to mind Mark Twain's amused reaction to Australian history. During part of his last round-the-world lecture tour, Twain visited Australia for almost four months in 1895. He had this to say about the country:

> Australian history is almost always picturesque; indeed, it is so curious and strange, that it is itself the chiefest novelty the country has to offer, and so it pushes the other novelties into second and third place. It does not read like history, but like the most beautiful lies. It is full of surprises, and adventures, and incongruities, and contradictions, and incredibilities; but they are all true, they all happened.
>
> [*Following the Equator*]

William Chamberlain's adventures 'all happened' too. However, in the absence of substantial first-hand documentation, the unravelling of his life story has had to draw upon a variety of secondary sources.

My heartfelt reaction to his [imagined] homecoming remained with me as I researched and discovered further facts in his later life. It was then that a closing personal tragedy emerged as the impetus for my narrator, William Chamberlain himself, to tell his story in a desperate effort to come to terms with his grief.

The Devil and the Deep Blue Sea is thus told in Chamberlain's 'own' voice as I have reconstructed and imagined his tumultuous life. And truly this is a life which validates George Eliot's famous epigram: 'It is part of probability that many improbable things will happen.' [*Daniel Deronda*]

Make this voyage with William Chamberlain and you too will surely be 'taken away'.

THE LATE CAPTAIN WILLIAM CHAMBERLAIN
The World's News, 3rd October 1903

PROLOGUE: THE DEVIL IN SYDNEY COVE

Contrary events in the natural world can set our mind to thoughts of fate and destiny ruling our actions. Are you superstitious enough to find portents and omens in your everyday life? My own father would always contend that we struck out into the unknown as free agents put on earth to rule over God's creatures. 'All things are ordered for the good and for the advantage of Man,' he would proclaim. 'We are delegated with a supreme authority over the animal creation and while we must not destroy them in wantonness, or in waste, we are at full liberty to convert them to our own purposes.'

By the time I was a grown man with my own sons I would interpret the killing flight of the harpoon, as it pierced the very colossus of all creation, as proud evidence of my preordained status: of man created to subdue the earth and have dominion. Now I ask myself if it was this pride that was struck down when my youngest boy lay in the bush with his throat cut, and worse . . .

For many days thereafter, my grief interpreted this lifetime of slaughter on the seas as being repaid by this bitter blow. Could I discern God's mocking hand behind this terrible loss? There was a time, already fading, when whaling fleets did whiten the Pacific with their canvas. Then I wholeheartedly believed that whether viewed as a nursery of bold and hardy seamen, or as a productive employment of capital, or as furnishing articles of necessity to human comfort, the whale-fishery was a noble enterprise. Now I confess that when I consider the habits of the great whale, how quietly it grazes through the pasture-ground which God has ordained for it and fitted so well to be its home, and when I observe the hazards that have to be encountered and the perils to be surmounted in its capture, and couple with this a consideration of the various other sources from which the human family can now be supplied with oil, I am belatedly doubtful about both the lawfulness and the expediency of the war we wage against the whale.

Enticed by the riches that come from vanquishing whales, we have disturbed the peace of their vast wilderness, violated their haven, wiped out all those unable to steal away to the wastelands of icy polar seas. And so this giant of giants has fallen prey to our weaponry. Since man shall never change, only when they cease to exist shall these enormous species cease to be the victims of our self-interest. They flee before our battlelines, but it is no use; our resourcefulness transports us to the ends of the earth. Their persecuted tribes have been chased so often and pursued so relentlessly from haunt to haunt that defeat and death is now their only refuge. Between some very few natural foes and its predatory human enemy, the great mammoth of the ocean seems doomed to extinction.

Such bitter self-loathing took me back to harrowing years of my own when I was a young boy and confronted a whaling captain who had suffered his own terrible loss. Did I come close to a brutal death when I was a child, see this same captain and his brother struck through like whales themselves, survive all this only to be saved for this final cruelty today? My life itself has sprung a leak and I am utterly dismasted. My hull is so deep-loaded with grief that I fear my own sanity is about to be cast adrift. I flounder in dark waters.

There is none of us whalemen who has been long in the service but has been maimed and broken in some way, and snatched often from the gaping jaws of destruction. I can tell of marvellous escapes and providential deliverances from the very threat of death that make you think a whaler, of all men, ought to be living under a ready apprehension of sudden and personal tragedy.

One would naturally think that a man's constant exposure to imminent death would impart a serious turn to his mind and induce a thoughtful regard to his latter end. But it is now a long time since the practical observation of men has taught me that familiarity with danger and death seldom produces a softening effect. Instead, it induces a moral hardness and effrontery that steel the mind against lessons of mortality and shed an ominous gloom upon the prospects of the soul.

I have talked with a good many whalemen and common sailors and have observed the conduct of irreligious men in times of battle and in more than ordinary dangers. Seldom, however, have I met with one permanently reformed and brought to repentance by seeing others drowned and die before his eyes and by what would seem to be the natural consideration of danger in his own case. So true it is, in the words of the preacher: 'The heart of the sons of men is full of evil; madness is in their hearts while they live, and after that they go to the dead. As the fishes that are taken in an evil net, and as the birds that are caught in the snare, so are the sons of men snared in an evil time, when it cometh suddenly upon them.'

And so it is. As fish in an evil net I have been snared.

For those who see a governing hand in the mere fall of a number, stop and consider this. Like a knell sounding over my childhood years the number thirteen rang out, sometimes even at my own urging. 'Thirteen years before you were born' was the pregnant phrase that started the oft-repeated yet endlessly fascinating story from John Wilkins, a marine private who had arrived on the First Fleet to New South Wales in 1788. He was my father's oldest friend and had written himself into the earliest records of the convict colony. The fabulous words from more than one journal had so frequently been quoted to him that their more educated vocabulary had entered John's own rough narrative. In this way, part of the litany of my childhood started not with 'Once upon a time' but rather with 'Thirteen years before you were born'.

'Thirteen years before you were born I rode on the back of a whale. Four of us were on the harbour fishing, trying to catch something, anything, to eat. A supply ship from England had been wrecked before it reached us and the colony was starving. Already they had sent almost half of us away to distant Norfolk Island so that the stores would not be burdened any further. But all we caught that day was the Devil himself. Death rose out of the depths, death like you have never seen. And once he set his cursed eye upon us the whale took three of my fellow marines into eternity.'

John would always pause here. The role of a Jonah is a thirsty one and he would only resume the tale once his glass had been refilled.

'This devil, a spermaceti whale, had been sighted in the harbour earlier, in July 1790. Those who were best acquainted with the nature of this enormous creature were not without a degree of alarm, which afterwards proved to be but too well founded. He differs, it seems, from his dull kindred of the North seas, and in nothing more than in a disposition to mischief. Some of the marines gallantly launched boats and pursued the monster, but to no avail. They returned with the sole success of having escaped him. He continued to range through the harbour, and was shortly the cause of a very melancholy event. Perhaps the taunts and the pathetic barbs they threw in his wake were repaid a hundredfold the following week? Even though we had already seen sights enough to turn hardened marines into shivering gills, this was a new wonder. But . . .'

And here the script called for an interruption from the audience. Once again we would call for a retelling of those earliest days in this land where summer in December, black swans and an alien landscape were only a hint of all that was strange and disturbing to British minds. And thus, encouraged by our pleas, the story continued.

'We had been warned; there were signs enough. Skies so threatening they spat lightning that killed our precious cattle and a rain of hailstones big as your fist! There were even wavering lights in the sky itself. Surgeon White called this the "Aurora Australis". The heavens flamed and some of the men seemed to have caught a fever of the brain from its appearance. Later, they swore they saw a monstrous snake dragging its coils through the camp at night. And some who walked off into the bush got lost and never did return!

'We were not welcome in this new country. The land itself even tried to shrug us off with a quake which turned the very ground beneath our feet into a churning sea! All this happened in the middle of our very first year as we clung to this coastline like barnacles to a rock. One of the convicts, Edward Cormick, who had absconded into the bush, was so terrified by this quake that he returned to the camp. He had been brave enough, or foolhardy enough, to face the Natives — "Indians" we called them — but this quake was more fearful than anything he had imagined. Even though he knew he would be hanged for his few brief weeks of freedom, the thought of another night alone in the bush, riding it like the back of a shaggy beast trying to throw him off, was more fearful still.'

Nothing could have brought such a disparate gathering together more than these stories of suffering and terror. They made us realise the precarious hold we had on life, that we were no more than a speck on this vast uncharted land. We were half a world away from England. Would anyone know, or care, if the bush and the savages and the darkness engulfed us? I would lie in bed at night and watch Father snuff out my precious candle and think how easily the flame of our small lives could be extinguished without leaving any record of our being. The bush would reclaim our pathetic scratchings and the Natives walk over what we had labelled touches of civilisation…

But John Wilkins was moving towards the climax of his story, the defining moment of his life itself.

'The monstrous creature no sooner espied our boat than he pursued and never left her till he had overturned us. At first he half swamped the boat with water. We pushed against the whale. So close was he to us that he filled all our horizon. His breath made us fit to swoon. In vain we threw out our hats, the bags for our provisions, even the fish we had caught. In these small ways we hoped to satisfy him or turn his attention. Our boat was twice on the Devil's back. I will never forget my sick fear as the whaleboat was suddenly lifted out of the water by the rising back of the whale, the hopelessly slipping motion as it shot down the heaving mountain. We were powerless. Then he tossed us into the sky and broke the boat all to pieces. After the boat was stove I rode the whale's back before I could even attempt to make for land. I clung to the whale like a sucking fish until a combing sea swept me off.

'None of the others could swim. With one sudden and tremendous gambol the monster consigned my three companions to their hapless fate. Then, although it was upwards of a mile, I got safe to the south shore.'

Here John would pause, as if to give thanks for his deliverance, then continue: 'I proceeded to the lookout house with the news of such a sad catastrophe.' And, as a final comment on the horror, he would add his frightening conclusion: 'The monster seem'd bent on our destruction and wreaked a terrible revenge.'

When I was a young boy myself I welcomed a contrary wind which filled my sails and made my spirits soar. My earliest memories are of being aloft in the eyrie which perched atop my father's house. Here I could imagine I was aboard a boat, that my flapping shirt was a vast outbellying sail and I was a captain searching the horizon for whales. Here I could be safely buffeted by the wind while my parents' strong hands steadied my infant exuberance.

My father was a bricklayer and had built this house himself from his own fanciful design, which included such an eccentric vantage point. Mother had her sewing room and its inglenook but Father liked to stand in his eyrie, as if on the prow of a ship, where he could clear the brickdust from his eyes. Here he would scan the Harbour and scour the rasp out of his throat as he drank in gusting draughts of sea air. Is it any wonder, then, that our Sydney-town neighbours soon labelled him 'Captain' Chamberlain? And what could be more natural than for his only son to assume that very title later in life through actual deed and training?

The love of adventure and enterprise that so soon discovers itself in an active boy seems to prompt the first inclination for sea; a longing curiosity keeps it alive, and nothing but a voyage will at last satisfy the youthful Argonaut, to which some parents consent in the hope that a life of danger and toil will soon sicken the romantic sailor and make him wish to live at home. This, however, seldom happens, and the first voyage casts the die for a future life at sea for the young adventurer.

It is somewhat remarkable that even boys in inland towns should often show this early desire of going to sea. I have seen it reveal itself there in a very romantic manner and terminate in an elopement purposely to embark. Among such boys the history of an old sailor is accounted the finest piece of eloquence and, whenever he appears, the subsequent narration of his voyages, battles and shipwrecks is listened to with rapture. The voyages of Drake and Cook and Commodore Byron around the world are famous in this way and eagerly read by schoolboys; but Robinson Crusoe has made far more proselytes to these kinds of adventures than all other mariners. His story from first to last is so full of incident, and in all his difficulties he shows so much courage, address and ingenuity, that the young reader will wander for ever in quest of just such an island as Crusoe's.

Certainly I was no different from these idle boys dreaming of adventures under sail and was a rapt listener to any old sailor's stirring narrative. My young imagination was often wound up to the highest pitch of excitement at the idea of exploring the trackless ocean; and to me the prospect of treading the quarterdeck as the ruling spirit of some stately craft seemed the very acme of all earthly ambition.

There were opportunities enough in Sydney Cove of boldly boarding whaling vessels and observing sailors whom I regarded as the finest fellows in the world. Now, a lifetime later, I realise that whales, imaginatively at least, swam into my ken and inspired the dreams of my earliest childhood. One of my fondest memories of those times is of a day when my father took me across the harbour to a point nearly opposite Dawes's Battery. The surprise purpose of this excursion was only revealed to me when, after an arduous climb, we arrived at an ancient rock face. To my astonished gaze an extensive carving was revealed on a sandstone rock platform — a carving of a whale, about the natural size.

The whole was remarkably correct in shape and its execution not altogether deficient in spirit. Its lines were very shallow and, but for the fact that it was the work of people entirely unacquainted with the use of metals, its execution might not have been remarkable. But, although shallow, the carving was cut on such hard rock that it was difficult to understand how people without iron tools could have produced it, and more difficult still to believe that it was due to mere whim or caprice.

With such tools as the Aborigines possessed it must have been a work of considerable labour and therefore undertaken for a set purpose and to serve a particular object. But what that object was it is impossible to say; probably it was a superstitious one. Sadly, over thirty years later when I returned to this same site with my youngest son Henry, it was to find that the rock had been quarried for building purposes.

Also in my own childhood there was the added inducement offered by a near neighbour of ours. This superannuated sailor was a source of the highest gratification. To hear him relate, with that quaintness of narration peculiar to his profession, the wonderful stories of his voyages, his adventures in foreign lands and the many thrilling incidents which abound in the jovial life of a sailor, was even more compelling than reading of such things.

After adroitly contrasting continual novelty and varied scenes with the dull monotony of a life spent on shore, in any occupation, he would conclude with a heartfelt wish which was a deadly blow to all the dissuasive arguments urged by my relatives. 'Oh, if I had twenty sons,' he would declare, 'I would make them all sailors.'

And thus my imagination was stirred and my young eyes were always straining to see beyond our Sydney-town horizon as my shirt-sail ballooned with air. My whole being yearned towards the open sea, towards independence and a life of continual variety. Now, after so long a life afloat, I believe that I can justifiably claim the tribute that is due to the whaler, that he might appear to the world in his true position. By the uninitiated he is too often regarded as ignorant and of a degraded caste, whereas in fact his occupation is often of extreme anxiety and trial, requires considerable fortitude and resolve for its proper performance and is, without exception, the noblest branch of our merchant-navy. I make this claim because, in the absence of war, it is the whaler who is best adapted to display the courage, perseverance and enterprising spirit of seamen.

This then is the calling I have followed. Even if my first voyage was hardly voluntary, began at too tender an age and was fraught with far more than the usual offering of peril and suffering, still I now account it a baptism of fire. A lifetime later, as I know myself to be sailing towards my final horizon, I am content to set my jib for family and friends. Increasingly, I find myself thinking of the strong currents and the wild winds which have carried me through almost threescore years.

Only last winter, almost seventy years after John Wilkins rode the whale's back, I was sorely in need of an old man's solace to turn my thoughts away from Henry's brutal murder. Thus I spent my days immersed in a wondrous book only newly arrived in the Colony. I have been told that it is but the imaginings of someone skilled with the pen. But surely this Herman Melville has hunted the whale in real life? As a captain myself, I cannot doubt his words nor the awesome respect he brings to his subject.

This *Moby Dick* of his is itself a whale of a book. Few things in its immense sweep pleased me more than the unexpected praise which its author gave to whalemen, praise for something which is not acknowledged frequently enough. This is to credit the part we played in charting and advancing this country of ours — not the landed part, of course, but rather the equally challenging watery realm that surrounds this greatest of islands. Melville called Australia 'that great America on the other side of the sphere … given to the enlightened world by the whaleman'. His informed compliments lifted my spirits momentarily when his praise continued:

> After its first blunder-born discovery by a Dutchman, all other ships long shunned those shores as pestiferously barbarous; but the whale-ship touched there. The whale-ship

> is the true mother of that now mighty colony. Moreover, in the infancy of the first Australian settlement, the emigrants were several times saved from starvation by the benevolent biscuit of the whale-ship luckily dropping an anchor in their waters.

Further support to Melville's case is surely provided by the fact that Cetus is the constellation in our own southern skies, a constellation which gives its name to the mammalian order Cetacea, which proudly delineates the whale and all its kin. Truly, one hears of writers that rise and swell with their subject but what unsurpassed magnitude has this author assumed in writing of Leviathan? In the impetus of putting words to paper, this Melville has surely proved that the heart can speed up with anticipation just as it does during the chase after a whale.

Among the many truths in his book, only one thing was more compelling to me than his description of the vengeful white whale; namely, the fearful portrait he drew of the insane Captain Ahab. When Melville wrote of him that 'on the passage home, he was a little out of his mind', I was much afflicted. Instantly I was transported back to a time when I was not quite eight years old, to late spring in the year 1811. It was then that the stars reached an evil conjunction and launched me on my life's stormiest sea. In that cruel time I first met the whaling captain of the *Frederick*, Alexander Bodie. It was as if Melville had recast the crazed Bodie into the heaven-storming Captain Ahab, yoked his madness to John Wilkins' true account of the vengeful whale and thus created his *Moby Dick*. However could this be?

But I overreach myself. This *Moby Dick* has been a delight and yet a sorry visitation. While it took me to sea once again, its dreadful conclusion reawakened such tumult and pain in my breast that I was for some days near kin to Captain Ahab myself in a frenzy of recovered emotions. From that overwhelming sea I must, for my wife's sake and for my dear children who yet remain, grasp one memory, one lifeline to sanity. And, indeed, nothing is more soothing to me than the recollection of my homecoming as a thirteen year old in 1817. For if I have learned one thing from my long and stormy existence it is this: that we die and are reborn several times throughout our voyage through life.

Part One

ASSISTANCE & PROTECTION

It is an early morning in 1817. The convict transport ship the *Morley* has been my home and my salvation for close on four months. I sail not as a felon but as a guest of the Navy Board in Portsmouth, whose honourable discharge has seen me safely consigned into the care of Captain Robert Brown. How different is my lot from that of these poor exiles below decks. They face a future torn from family, friends and home, with no hope of a reunion; I anticipate a family who will welcome my return as if it were from the grave. For surely they believe me long dead, so sudden and mysterious was my disappearance.

The convicts sail with an introduction to the Governor of the penal colony of New South Wales: their indent papers which briefly list their paltry characteristics, their crimes and their sentences. Most are consigned to a term of seven or fourteen years of penal exile in Botany Bay. I too sail with papers which are also for the eye of the Governor, but their text is one of hope and comfort and promise at the end of this voyage.

My letter, like myself, is entrusted to the safe keeping of Captain Brown and comes from London's Downing Street. It has been written by the Admiralty's Under Secretary and is addressed to His Excellency, Governor Macquarie.

> Sir,
> It having been represented to Lord Bathurst by Dr Quarrier, Surgeon of His Majesty's Ship *Leander*, that William Chamberlain was forcibly taken away from Port Jackson about four years back in the *Frederick*, South Sea Whaler, but that he is now desirous of returning to his Friends in New South Wales, I am directed by his Lordship to enclose for your Information Copies of Two Papers relating to the circumstances of the Case, and I am to desire, as he will proceed in one of the Convict Ships now under Orders for the Colony under your Command, that you will afford him all the Assistance and Protection in your power on his arrival at Port Jackson . . .

Standing on deck, I strain to discern anything, for we are supposedly well along the east coast of Australia and headed towards Sydney Cove. Behind me a voice comes out of the darkness: 'William Chamberlain, is that you, my boy?' It is the voice of the Captain himself. For him to be on deck this early in the morning can only mean one thing: that we are much closer to our destination than I imagine. 'It is still too dark and misty to see anything like the shores of your home yet,' he says with some sympathy. 'And, anyway, we cannot breach the Heads till the pilot arrives.' So, at least I now know that we are hove to off the very entrance to the Harbour itself. Even so, not even my keen eyes nor my most heartfelt desire can descry anything.

In this part of the world it is by no means uncommon for the wind to change its direction suddenly and violently. As yet, there is no towering lighthouse built on the south head of the Harbour to serve as our guide. Because of this, the orders 'Shorten sail!' and

'Heave to!' ring through the fog as the Captain takes precautions against being drawn onto the unseen shore. There is thus an oddly dreamlike quality to my homecoming: the world is leached of all clarity and definition and earth and water, air and sky all come together in a diffuse milky immateriality.

The *Morley* had left England in mid-December 1816 with 175 convicts aboard guarded by a detachment of the 46th Regiment. As usual, the sailors saved up their allowance of grog, half a gill, so that by Christmas Day any one of them could believe himself to be an Admiral. This occasion was always a general holiday and feast of Bacchus but, as someone more used to navy discipline, I found their intoxication astounding. And it seemed curiously hard that men so eager to get drunk should have been so urgently encouraged to drink and then so brutally punished for its effects thereafter.

The Surgeon on the *Morley,* Dr Espie, had solemnly promised Dr Quarrier to oversee my well-being, but I had paid little notice to the former, so fraught was I with my imminent leave-taking. And indeed, my farewell at Portsmouth had reduced me to a flood of tears. Quarrier, himself in tears, had smoothed my hair, all the while reminding me that someone who had served so bravely before the walls of Algiers must now summon up further courage.

I searched deep into his eyes and in return received the steady, penetrating gaze of the good Doctor. This was a final look and a parting memory to take with me for ever; a look and a memory which now had to last me a lifetime. What was I seeing that words and actions had so often told me? I knew at once. I was seeing the eyes of a true father; true in their emotional blessing and manly fortitude. I clung to that look and those emotions as a drowning sailor to a floating spar. And thus I consigned myself to a long voyage once again.

My sad reverie was broken only by the sight of the many poor wretches bound for penal servitude. As soon as they had been assembled, Dr Espie made ready to address them. I could not help comparing him with the only friend I had just left and my first opinion of this new Doctor was not charitable. In retrospect I can see that nobody could have measured up to Dr Quarrier, who had become as dear as a father to me. Indeed, I realised that his face was more easily and eagerly recalled than that of my own father back in Sydney-town. Was my family still alive? How would they have changed in the five years that I had been away? It was thoughts like these which haunted my voyage home.

At first meeting I had judged this Doctor Espie to be nothing but a versifying bore whose introductory address to the convicts was hardly likely to lift their spirits from thoughts of exile or to elevate their tarnished souls.

'The object of my calling you together', he began, 'is to make a few observations regarding the voyage which, under Providence, we are about to commence. On your good and correct behaviour during this voyage your future happiness will depend in an infinitely greater degree than, I apprehend, any of you can form the least conception. Let me, therefore, entreat your undivided attention and most serious consideration to what I am about to offer, as it materially concerns your own welfare.'

Already I could see that these felons needed no more sermons to remind them of their fate. Few even bothered to raise their heads and pay any attention to this latest offering, which surely only repeated sentiments they had heard endlessly before.

'It is far from my intention to add to your distress by upbraiding you with the errors of your past life. The word exile, or banishment, sounds harshly on the ear and must ever convey to the heart the most acute feeling of anguish. Our Native land, containing the scenes of youthful amusements and innocent pleasures, abandoned perhaps for ever! The dear ties by which nature had united you to your families, relations and friends must become lacerated; yes, to be compelled to separate from parents, children and wives, dear as life, and to be parted from them for ever is, it must be confessed, a truly bitter thought. Under this accumulated load of misery is there no ray of hope, no relief to be obtained? Yes, my unfortunate friends, not only relief but consummate happiness, even joys unspeakable, are within the reach of every one of you.'

Surely, I thought, these wretches all knew what they were leaving behind. And just what comfort did the Doctor expect them to receive from his words? But then, as I half listened, I realised that there was some solace to be found after all.

'The Holy Scriptures abound with merciful promises of forgiveness, and our blessed Redeemer addresses us all. His language is that of a tender father earnestly soliciting a lost child, on whom his fond affection is lavished, to return home.'

The pertinence and pathos of these familiar words struck me most forcibly. It was as if this sermon was particularly addressed to me and to my situation. And so, as we embarked, my spirits soared for the first time in weeks and I began to anticipate, fully, a homecoming and a family reunion.

And by the end of that voyage, I found that to be the only sermon inflicted on the convicts. After admonishing them, he seemed to consider this part of his duty entirely fulfilled; his sole occupation thereafter was in dispensing medical rather than spiritual succour. Not only my good health but that of the convicts was equally considered and it was a tribute to his care that not one died on our entire voyage.

EXTRACT OF A LETTER FROM DR ROBERT ESPIE TO DR DANIEL QUARRIER

To my esteemed Colleague

Dear Sir,
The tender trust with which you consigned young William Chamberlain to my care on his homeward journey to Australia, let alone the bonds of sincere amiability which have been generated during our voyage together, encourage me to write this letter. I have the pleasure to inform you of our safe arrival in Port Jackson, in New South Wales, 10th April 1817, after a swift passage of forty-one days from the Cape of Good Hope.

The boy has gained much on this voyage, both in stalwart good health and in patient fortitude. I cannot doubt that in restoring him to his family they will regain not only their son but one whose survival over near insurmountable odds has tempered his mettle far beyond his years. On the happy occasion of this reunion, I fully anticipate a most heartfelt scene. William's family cannot but believe that after so protracted an absence he was entirely lost to them for ever. His extraordinary return will be more in the nature of a resurrection than a reunion. I shall be a concerned onlooker at its unfolding, much gratified to have played a small role in the final scene of his moving drama.

During the four months that we have voyaged together, William has recounted the circuitous journey in both distance and emotion which he experienced. That he survived such vicissitudes at all and is now returned safely in glory is no small cause for congratulation. I know, Sir, that William's family will praise your name and give thanks for the compassionate care which you bestowed as a surrogate parent in their absence, once I fully apprise them of your role.

That William will make a mark for himself in the developing colonial society of New South Wales I have no doubt, although it will have to be in a field which offers exertion, adventure and enterprise sufficient to satisfy his recent triumphs. Further to the duties you charged me with, I can report that William has attended to his studies on the voyage out. For someone so experienced beyond his years this was not always a welcome task, as there were plenty of novel distractions. These included a fearful storm and, shortly thereafter, the awesome spectacle of the Aurora Australis.

So far as medical matters are concerned, this has been a singularly healthy voyage and William has benefited much from his temperate days aboard without apprehension or danger. As Surgeon on the *Morley* I believe I can assert that on this voyage there was an abundant supply of medicines and that the different articles were of the very best quality, having been furnished from Apothecaries' Hall itself. Indeed, as far as a solicitude for the health of this ship was concerned, I am persuaded that if a number of respectable families had embarked as colonist-adventurers, the outfit for the prevention and remedy of disease could not have been more complete and efficient.

On coming on board, each convict was supplied with a new suit of clothes together with a change of linen. From former experience I was fully acquainted with the necessity of a scrupulous and unremitting attention to cleanliness and ventilation; for which purpose the whole of their bedding was daily brought on deck to be aired. The prison which contained them was daily washed and thoroughly cleansed and, though it was the seat of their confinement, they were allowed, under a sufficient guard, to come on deck in parties consisting of a third of their number, where they continued in rotation about two hours at a time from six in the morning until evening approached.

Two fumigating lamps with vinegar were kept burning in the prison from morning until night; they had likewise wind-sails and air-pipes night and day to ventilate below decks. In addition to these precautions exercised under my daily superintendence, they were three times a week inspected jointly by the master and myself, at which muster each convict brought up his clothing for our examination.

Although their crimes had consigned them to a distant land, I felt that the amelioration of their lives was as much the wish of the legislature of our country as their separation from the main body was expedient to arrest the source of the contamination. It was necessary to guard them against revolt and consequent escape, which they hourly meditated, but I tried never to forget that they were men and women, nor abandoned hope that they might be reclaimed into good citizens.

Concerning the future, which it was my constant solicitude to impress on their minds, they appeared, however, to be wholly indifferent. Many did not believe in the doctrine and others were so heedless of its consequences that it served only as the subject of vulgar jest or impious derision. This apathy I can only attribute to want of reflection. When their recollection reverted to former scenes it was instances of impure gratification or of successful plunder which were recounted with triumph by the authors and always received the applause of the hearers.

Situated as they were between the depths of the ocean and the highest reaches of the heavens above, one would have thought that their minds would turn naturally to the Author of these wonders. But I have never seen men so unmoved by the beauty and magnitude of creation; here they were, quite mindless of His all-seeing eye.

Not long after we had crossed the Equator, their utter insensibility to anything other than brutish sentiments was clearly demonstrated. Not even the wonders of a phosphorescent sea could rouse their dulled senses. When we were about seven degrees to the southward, the tropical waters seemed aflame at night. Around the bow, along the side and in the wake of the *Morley* the sea was on fire, as if agitated by our progress over the depths. Even in our latitude at home, seawater phorphoresces beautifully when agitated; but this exhibition is faint in comparison with what we saw in these tropical regions. Brilliant white flashes and globes of light of great size illuminated the ship's path. Occasionally a wave would break at a distance from the vessel and from its crest issued a ball of light that seemed to play upon the waters like the flickering rays of the 'ignis fatuus'. Porpoises and other creatures sporting around the ship would leave a luminous train after them, winding frequently in serpentine curves.

I fetched William from his sleep on more than one night to witness this uplifting spectacle. He is a sensitive boy and also noticed, and was disappointed by, the convicts' seeming insensibility to this and other wonders of creation. He commented that even the common sailors aboard took time to revel in this nightly spectacle. One fellow, untutored but obviously struck with admiration, solemnly explained the phenomenon to William.

'Well now, young fellow,' he began, in reaction to William's exclamations of delight and wonder, 'that is what we call a River of Milk. It is an example of the great bounty of nature: the spawn or seed of whales that abound in these waters. See how it outshines even the stars above?' And thus with this engaging story, our simple poet added a further unforgettable dimension to our experience.

I drank into my innermost memory this lesson of the power and grandeur of creation. These lights are surely revelations of the teeming life which crowds every foot of this vast ocean surface. In the absence of knowledge to explain such wonders, I am content and happy to look on and admire and simply accept that 'Such are God's ways on deep water'.

When we were off the Cape of Good Hope a very heavy squall came on during the night. There was a general apprehension that the vessel could not long withstand its fury. This appeared to me to be a favourable opportunity to impress the minds of the convicts with a due sense of their awful situation. As well as I was able from my own apprehensions, I endeavoured to exhort them to employ the short time that probably remained in prayer and repentance — but all in vain. The violence of the tempest inspired them with additional excitement and my admonitions were drowned in a roar of blasphemy. They recollected their former debauchery at country fairs and began a song commemorating their licentiousness and compared the rolling and pitching of the vessel to the drunken orgies which they enjoyed during those festivals.

The violence of this storm drove us much further to the south than we would otherwise have voyaged, almost to Macquarie Island. But it was because of this, no doubt, that we were entertained by the pyrotechnical brilliance of the aurora. It was first seen as a diffused light over the southern horizon above a heavy bank of cumulus, the sky being perfectly clear elsewhere and the stars shining brightly. Out of this tremulous luminescence a streamer of light began to move towards us. Soon separate and distinct beams flashed in every direction — horizontally, vertically and obliquely. Most astounding of all, we then witnessed cones and circles of light which travelled rapidly over the whole sky.

I was instantly reminded of our esteemed and still lamented Captain Cook, whose second foray into the Pacific had revealed the wonders of what he called the 'Southern Lights'. In February 1773 he had recorded the spectacle as superior to the Aurora Borealis, with colours finer and flashes more quick and beautiful. And certainly, as I watched entranced, a remarkable change occurred. An arch of bright green light fading off into yellow formed over the southern horizon and rose rapidly to a higher and higher altitude. This was followed by similar shapes in regular succession until there were six quite distinct arches marching across our vision. All of them were bright green and yellow at the tops and of a rosy hue where they touched the horizon.

About an hour after the moon rose another remarkable change occurred. Extraordinarily vivid coruscations rolled over our heads in beautiful columns, then as suddenly changed into the fringe of a curtain or shot across the hemisphere like a serpent. Light coursed down like the strings of a harp and I distinctly felt the hairs on the back of my neck stand up. Whether this effect was due to the lights or was my own awestruck reaction I cannot say.

In such an extensive and ever changing panorama it was impossible for one observer to accurately describe the many interesting aspects which so continually presented themselves and so rapidly vanished. Each one would catch a glimpse of something different and could only form a very general idea of what actually took place across the whole immense canvas of the heavens. Although the *Morley* was running with a smart breeze and much beset, the crew could scarcely be kept from this entertaining spectacle above instead of attending to our course.

It is as true of these lights as of many other things of which we have no sure knowledge, that thoughtful men will form opinions and will make such guesses as seem reasonable. But amongst the convicts the spectacle was the cause of much alarm, since they took it for a heavenly warning. Some few of those ignorant souls on deck were even apprehensive lest its sensuous fires burn their flesh or singe their hair!

The vulgar, ever prone to superstition, judged the lights to be a kind of prodigy and the forerunner of battles, famine, plague and pestilence. They even professed to see in the sudden coruscations brandished spears, marshalled armies and plains drenched with blood! If there are mystic signs and portents in these lights, however, then it is a kind of golden writing which we do not have the key to decipher. William, who had witnessed the bombardment of Algiers at night, understandably felt that this was a much more peaceful celestial spectacle and took it for a particular welcome back to his southern homeland. I could do no better than quote his own words here: 'To see this glory in the skies you would think that you were looking into the gates of the New Jerusalem.'

To return to those below decks.

I discovered that the greater part of these convicts had been early initiated into the mysteries and practice of their profession and that the daily pursuit of their craft was not likely to be discontinued by abstract precepts. However these persons might differ in the complexions of vice, there was one strong feature which pervaded the whole mass: the pertinacious and unflinching assertion of falsehood and a solemn denial of the truth, even after its existence had been demonstrated. Although they might combine by strength or stratagem for a common plunder, yet these partners would rob each other of their dividend of the booty with as little ceremony as they had defrauded the original owner.

The first difficulty I had to encounter was the want of reciprocal communication: they comprehended sufficiently everything I said to them, but in their discourse with each other they spoke a language wholly unknown to me. This copious and figurative vocabulary I was compelled to acquire and for my progress in this new tongue was principally indebted to a convict who had been a schoolmaster. By this acquisition I was in a great measure enabled to ascertain their proceedings and comprehend their plans.

From enquiries in which I could not be mistaken, I found that most of the men had enjoyed a premature commerce with the other sex. From such intimacy they became more thoroughly corrupted and incited to the repeated commission of crime for the maintenance of female favour. Of their own degraded condition they appeared to be wholly insensible and this moral insensibility was manifested even towards their nearest relatives, and of this fact the following is a striking instance.

When the *Morley* lay off the English coast, a convict about thirty-five years of age, who appeared to be one of the most orderly and decent of his comrades, was visited by his wife who had travelled from London to take affectionate leave of her husband. She had kindly brought with her some articles which she conceived might contribute to his comfort during the voyage. After a tender and distressing adieu she departed and had nearly reached the shore when I perceived the boat returning. She had discovered that in their parting embrace he had picked her pocket of the money which was to convey her to town, and during the final pressure of her hand he had also dexterously contrived to detach the very wedding ring from her finger. It was useless to compel him to restore the property. He reiterated the most solemn protestations that he was innocent and could not even be prevailed on to see his wife again — under the hypocritical pretence that his feelings were so lacerated by the parting scene that he could not sustain a repetition of the conflict.

Their hypocrisy was inscrutable and there were not wanting a few who performed to admiration the mimicry of repentance. I confess my own gulling by their charades. Though my hopes diminished as I became acquainted with them, yet there was an instance during the voyage of correct conduct in a young man which afforded me some consolation for their general failure. This youth had received a competent education and without being a traitor to his community he appeared to be weaned from their practices. Although obliged to associate with the others he never joined in their indecencies. He officiated as clerk during Divine Service and regularly distributed the Prayer Books. He exerted a proper spirit of activity and diligence in keeping the prison clean and his continuance of exemplary conduct could have recommended him to the Governor. On our near arrival at Port Jackson, however, when the convicts were assembled for inspection, this hopeful penitent was detected in picking the pocket of another . . .

THOUGH LOST, YET HE IS FOUND

Being just thirteen years of age, I imagined that I was the youngest on board the *Morley*. However, my observations of the convicts who were daily brought on deck for healthy exercise and fresh air soon revealed that among their number there were at least two boys younger than myself. It is a sad and sobering thought that, as I journeyed home to resume a life among family and loved ones, these other boys, already well schooled in infamy, were leaving such comforts behind them, perhaps for ever.

My deeply felt expectation of being reunited with my family was in dreadful contrast to the exile our convict passengers were facing. After my enforced separation from family for five years I was ready to commiserate with any felon who had virtually no hope of ever seeing homeland or family again. But I will admit that some of my fine sentiments were forgotten at the first sight I was given of the women convicts as they came on board.

As much as I yearned for a parent's fond embrace once again, there was little evidence of maternal or even of feminine attributes in these poor wretches to remind me that they were kin to my own dear mother. Their arrival on the ship was anything but promising. That greatest ornament of women, their hair, had been cut close and rough and they embarked with a great deal of crying and complaint. Their abusive language and complete lack of personal dignity were a cruel affront to their sex.

But, as they reconciled themselves to their exile and as they recovered from their seasickness, I found that my feelings of disgust and abhorrence were wearing off. There was, I decided, not so much difference between them and the rest of the world after all. The sailors soon became intimately acquainted with them and I was frequently employed to write love letters for the old salts.

Groups of the women were allowed on deck during the day and in fine weather some even had their meals there, but all were sent below and locked up at sunset. The system of government which operated on board was founded on the old adage, 'Set a thief to catch a thief'. Several of them were promoted and promised rewards to watch over the rest — thus, one not being able to trust the other, they were all kept under. These constables were possibly far more tyrannical to their fellow prisoners than free ones would have been. They had several fights among themselves and once, alarmingly, there was a general mêlée caused by the tyranny of these constables.

This made more noise than all the fiends of Bedlam let loose. Our Captain's response was to put the hatches on. When this failed to quell the riot, however, one enterprising liar told them that the crew would leave in boats and scuttle the ship if peace was not restored. But on this lie getting wind they raised an even more formidable noise. Imagine the lungs of so many women combined in a strident chorus — it was as dreadful and as upsetting as anything I had experienced at Algiers. Peace was only restored when Dr Espie discovered their dire fear and exposed it for the cruel fiction which it was.

Most of the women were soon quieted and officers were sent down to apprehend the last and most violent, who were taken on the poop and put in irons. The mate, who expressed himself taunted beyond endurance by their abuse, was violent to several of them there. When he tried to gag one of these unfortunates he was served out when she caught his fingers between her teeth and bit them to the bone. Thereupon he struck her violently in the face with his disengaged hand. At this unmanly act there was a great shout among the women. It was a mark of our Captain's forthright justice that, when an enquiry was made into the cause, the mate was severely reprimanded.

Even with these disturbances, the *Morley* finally made a record-breaking run. But, oh, to have come so far and so quickly and now at this very last minute to be delayed by a fog! This was not the seafaring I had already experienced. After all, even the smoke of battle had not deterred another Captain from sailing us into dangerous waters only the year before. But this was Captain Brown's first voyage to Sydney Cove and this final feat of navigation was, with good sense, only to be undertaken in clearer weather.

Impatiently, I recalled my answers to the Navy Board in Portsmouth not long before I was given passage on the *Morley*.

'How long is it since you left Port Jackson?'
'I believe it is rather more than four years.'
'Would you wish to return to your Father, if we can obtain a Passage for you?'
'Yes, Sir.'

I had already seen enough of death and destruction to wonder if my parents were even still alive to welcome me home. I knew that it was more than this clinging fog which made me tremble. If only I could discern something familiar, something which convinced me that I was, indeed, within landfall of home and family. But all around was this swirling white pall. 'Only see,' cried another more keen-eyed than myself, 'there, over there.' And he pointed triumphantly to a dark shape louring out of the dim light. My anguish to recognise any familiar landmark shook me anew until I heard a calming voice: 'Easy, lad, that's surely the South Head of the Cove we're seeing. With dawn breaking and a fair wind we'll soon be delivering you like Lazaras himself to the Governor.'

I peered forward into the breaking light and clearing fog. Was this really home? I recalled standing at tense lookout on other decks while it was death that came at us across the water. Then it was a pirate crew who broached our darkness and, much later, the threatening ramparts of Algiers. But that was past, all past, and my stormy sea had been crossed. The Captain ordered the ship forward and other eyes as keen as mine, although nowhere near as passionate, guided us between the Heads as if through the gaping jaws of a Leviathan itself.

We arrived in Sydney Cove on 10th April 1817. Now I was faced with another leave-taking, saying farewell to the other discharged soldiers who had accompanied me home. The early morning fog began to lift and by the time we were safely anchored in seven fathoms in Sydney Cove I had a clear view of the town. Although it had been only five years since I'd left, there seemed to have been many changes. My eyes darted everywhere, frantically searching for some familiar landmark or building. I was close to tears when a caring hand on my shoulder from Dr Espie told me that my agitation was not unnoticed. His sympathy was immediate and his advice both welcome and effective.

'Close your eyes, William, close your eyes. Be calm, breathe deeply and listen, listen carefully. Keep your eyes shut and tell me what is familiar.'

Reluctantly I did as the Doctor advised and, without the distraction of sight, there were suddenly sounds and smells evocative enough to convince those senses at least that I was truly home.

Most familiar of all, there was a raucous peal of laughter from the nearby shore. Here was a welcome indeed! No bird of ill omen this time, like the one that had circled the *Frederick* and sent her Captain and his brother to a cruel fate. This bird, this laughing jackass, broke through my turbulent thoughts and could only raise my spirits in unison with its unexpected hilarity. And then another sound arose, equally poignant and welcoming. The peal of Sunday morning bells rang across the harbour in a second greeting and instantly I was reminded of the day, last October, when bells celebrated our Navy's arrival into Portsmouth. Surely these were auspicious omens?

After we announced our arrival with a fifteen-gun salute to His Excellency, Governor Macquarie, I finally set foot ashore. But even now I was not allowed to proceed in search of my family. In company with Dr Espie I took the letters and papers that had come from Downing Street to Government House, fulfilling my final act of naval duty before resuming civilian life.

As we made our way through the early morning throng around the Cove I eagerly scanned every face we passed. Maybe one would be familiar to me; maybe someone would recognise mine? No doubt I had changed in five years. While I had certainly grown taller I had not really begun to acquire that stature which later suited my own commanding role as a captain. My recent ordeal had also left me easily fatigued and I was sure that my pale face was a noticeable contrast to the tanned figures I saw all around the shore.

It was only a twenty-minute walk from where we had berthed to the sentry box at Government House. This was nothing like the magnificent Gothic villa which would later be built on Bennelong Point but more a jumble of parts. It lay very close to the town at its eastern extremity. After stating our business to the guard we were told that public access was at the back, where we finally presented ourselves to the Governor's Secretary. He dutifully took my naval papers, Dr Quarrier's detailed Account and our personal particulars. And that was all.

I certainly had not expected an audience with the Governor himself, nor an invitation to step inside his imposing residence. But as we retraced our steps I realised that one part of my life had just quietly finished. This last duty to be performed under the Navy's orders had hardly been heralded by cannon fire or acknowledged with stirring speeches. But while such a calm finale was the exact antithesis to my recent experiences, it was, I reflected, singularly appropriate to one now seeking a reunion within the quiet bosom of family life.

Now it was I who led the way back, increasingly certain of my steps as I recognised familiar streets and buildings among much that was novel. My breathing became short and shallow. I felt as tense as when I faced action beneath the ramparts of Algiers. An overwhelming emotion caused my heart to beat more rapidly. Like someone attending a stage performance I disjointedly observed myself and my reactions. I felt close to tears. My hand worried my trembling lip. Joy and fear were inextricably linked.

And then we were almost within reach of my home, or at least what I had called home five years before. Would my family still live there? In what state would I find them? Church bells were tolling again as we approached a small Presbyterian chapel where the service was about to begin.

I raised my eyes to its slender spire and silently prayed that my protection and safe deliverance were only a prelude to a long-anticipated reunion. It was in this calm moment, as I gathered inner strength, that a swooning scream wrenched me back to earth.

'Oh my God, William. My dear boy! William Chamberlain ...'

Then, with a choking cry, a woman fell fainting to the ground. My eyes took in the congregation, who had stopped filing into the church and who were now staring dumbfounded, turning their gaze from the woman's prostrate form to me. I could see questions in their eyes and as the Doctor knelt down to tend to this herald who had so dramatically announced my arrival, he asked, 'Is this your mother, William? The poor woman swooned when she set eyes on you.'

In great anxiety I knelt with him. But then my worst fears were realised, for when I gazed on the pale face lying before me I could not recall its features as my mother's! She had recognised me but her face meant nothing to me at all. Did I not even know my own mother?

I was aware of voices all around. It was the congregation who were whispering my name. To me their muted words sounded as loud as the broadsides at Algiers. Then I was aware that the crowd was moving aside and so I made to escape from this gaping throng. But as I stood up, someone enveloped me in an embrace and soothed me with a joyous repetition of my name. Although I was clasped so tightly that it was impossible to see, now at last I knew I was truly home. Here I clung, held so close by my mother and caressed by the sound of her voice, a voice I could never have forgotten. 'Sweet William, oh William, my sweet William ...' These were endearments so familiar, endearments which I had never expected to hear again. And then, yet another blessing, for over her shoulder ... surely that staring face now breaking into a trembling smile had to be my father!

There was much confusion thereafter and no doubt much that I have since forgotten. The poor woman who fainted was soon revived and proved to be a neighbour and close family friend. There were questions from all sides. Eventually, the minister came out into the middle of all this tumult to see why his congregation was not dutifully filling the pews. When the extraordinary reason was broadcast from a dozen excited witnesses, we were all exhorted to kneel. The service that morning was conducted outside where we knelt under the open sky, and the sermon composed on the moment had a short but moving message: 'Though dead he is alive again; though lost, yet he is found.'

AN ACCOUNT BY DR DANIEL QUARRIER

William Chamberlain's rescue from the *Frederick*, his subsequent life, and his recent heroic service on His Majesty's Ship *Leander*, Captain Edward Chetham, C.B., during the Bombardment of Algiers, 27th August 1816. Compiled by Dr Daniel Quarrier, Surgeon on the Frigate.

The following Account, supported by transcriptions from official records, is intended to tell that part of the remarkable story of William Chamberlain in which I myself have played a not inconsiderable part. In the five years that young William has been in my care I have taught him to read and write so that one day he may be capable, if so moved, to record his own tale in his own words. But as he is yet of tender years I have felt that another duty I can willingly accept is to chronicle his life thus far.

I feel that his scarcely believable record of juvenile suffering, piratical attack, naval battles and sundry adventures demands the imprimatur of someone like myself to give credibility to what might otherwise be deemed a young boy's fervent imaginings. I also fully acknowledge to myself that in the compilation of this Account my anxiety and grief at parting with the boy, even though he goes to rejoin his own father, will hopefully be assuaged by its clear message of bravery, charity and ultimate survival.

After Britain, by her perseverance and by the bravery and success of her Army and Navy, had succeeded in destroying the revolutionary power of Napoleon and in restoring the peace and independence of Europe, it was to be hoped that she would have had, at least for a considerable lapse of time, no cause to again appeal to arms. The contrary, however, was the case. Immediately after her Navy was reduced to its peacetime establishment, and everything bore the appearance of confirmed and permanent tranquillity, then an ancient foe reaffirmed their presence.

These were the Corsairs of North Africa, Muslim powers who lived by piracy. They hailed from that sinister place Barbary, which not only took its name from the Berbers — the indigenous inhabitants of the region — but also from 'barbarian', which is how Europe commonly viewed them. From the time that Englishmen had discovered that their real vocation was the sea, they had been fighting these pirates whose practice it was to enslave their Christian prisoners, accepting ransoms in respect of some. Although spoken in 1799, the great Horatio Nelson's retort to this insulting behaviour was widely reported and has not been forgotten. 'My blood boils,' he wrote, 'that I cannot chastise these pirates. They could not show themselves in the Mediterranean did not our country permit. Never let us talk of the cruelty of the African slave while we permit such a horrid war.'

The ravages of these detestable pirates were chiefly committed upon the vessels of the weaker powers, and particularly upon the defenceless coasts of Sicily, Sardinia, Calabria and the eastern side of

Italy; but no ports of the Mediterranean shores, except France, were free from their incursions. It had long been deemed an indelible disgrace to Europe that she permitted this piracy; nay, more, that she not only submitted to these things but also frequently sent presents to these piratical states.

Naturally, such of the European powers as had no Navy, or a Navy consisting only of a very few ships, could not be expected to do much against the Barbary powers; but the maritime powers might have been supposed and expected to exert themselves at least to protect their trade and their subjects from these states. After the end of the Napoleonic Wars, a new and energetic British policy towards slavery demanded a solution in the Mediterranean.

Even Britain had sent presents to these Corsairs, and exchanged courtesies, and thus demeaned and disgraced herself. No sooner had Napoleon been sent into final exile than this cloud again darkened Britain's peaceful horizon. Algiers was one of three main Barbary ports and its Turkish Corsairs, who had been the hereditary scourges of the Mediterranean for centuries, were next destined to feel our righteous anger. For over two centuries Britain had repeatedly sent squadrons to Algiers to confront them, but all our bombardments, let alone the natural justice of earthquakes and epidemics, had failed to eradicate the evil.

I cannot but wish that some of those English gentlemen who travelled in search of pleasure in the Mediterranean had paid Algiers a visit, even for one week. I am sure that they would not have failed to feel, like me, the degradation to which the Christian name was exposed; and then endeavoured, on their return home, to exert their abilities and influence in a cause which no one doubted to be meritorious, but which actual inspection would make every man feel to be a solemn, religious and moral duty.

The Christian slaves in Algiers were of all ranks and ages. Many had been distinguished in various military and civil capacities; and noblemen, scholars, churchmen and captured ship's crews all became companions in misfortune. And not men only, for women and children were also among the victims, in conditions so hideous and revolting that it seems almost miraculous that any human beings could have survived such sufferings. These slaves were of nearly every European nationality and, when their respective governments made only diplomatic efforts to secure their freedom, the Dey of Algiers not unnaturally assumed that he could continue to oppress them with impunity. Corruption, insolence and unscrupulous robbery were the gentler characteristics of this ferocious and contemptible government.

Behind his solid fortifications, the Dey counted upon the probability that internal dissensions among the European governments would prevent any serious and united effort for the abolition of slavery. Of England he had not the slightest fear; the ill-timed rhetoric of the peace-at-any-price zealots having led him to believe that British anti-slavery sentiments would be expressed merely in philanthropic circles or by diplomats armed only with verbal eloquence, unsupported by the irresistible logic of big guns, which was the only form of argument to which the cynical Turk was open. Ultimately, we stood foremost among the champions of Africa as opposers of the slave trade. But the miseries endured by the unhappy Negroes were not greater than those of the Italians, Greeks and Spaniards in Algiers.

The United States of America was the first to set an example of spirited resistance to the pretensions of these Corsairs. The new Dey had been proclaimed in April 1815 — his predecessor had ruled for only sixteen days before being slowly strangled to death in public by his own Janissaries. Truly, it has been said that rarely more than one in twelve of their Deys has the good fortune to die in his bed! In mid-1815 the Americans concluded a treaty with this new Dey and placed an American consul in that town. We were somewhat shamed by this initiative, coming as it did so soon after our Navy had been trounced by theirs in the war of 1812. Within a year, Britain began her own negotiations with the Barbary states.

As Commander-in-Chief of the Mediterranean Fleet, Admiral Exmouth endeavoured to negotiate the abolition of slavery in late March 1816, and even secured the release of many Christian prisoners. However, as soon as Exmouth's fleet sailed for home, the Algerines clearly showed their intractable nature. Further atrocities thereafter, including a massacre of Christians at Bona, near Algiers — of Christians who were under British protection — showed how clearly our peaceful negotiations would never achieve their desired goal. The Algerines had no intention of discontinuing the slave trade and, by the device of hoisting the British colours on some of their galleys, the new Dey decoyed on board and kidnapped others to replace those that had been released.

Scarcely had Admiral Exmouth arrived home than news of these latest atrocities aroused the Admiralty to offer him as large a force as he might think necessary to return and deliver justice. His past experience in bombardment was fortuitously coupled with a most recent and precise survey of Algerian fortifications. At the beginning of 1816 he had sent a party to reconnoitre the fortifications and had instructed the men precisely which particulars, including soundings, were most urgently required. He thus knew the Dey's resources as well as he knew the capabilities of his British seamen. He explained how a small force of line-of-battle ships could destroy the fortifications, while another could cover these ships from the fire of the batteries south of the town and frigates could deal with the batteries on the town walls.

Although our *Leander* under Rear-Admiral Milne was then on the point of sailing as flagship on the North America station, we were reassigned to Admiral Exmouth's squadron. Together with other ships, we joined them at Plymouth and Rear-Admiral Milne requested permission to remain with us rather than proceed to his station in Halifax. Accordingly, on Sunday 28th July 1816, shortly after midday, with an ebbing tide and a steady northerly wind, we formed up and set course. Accompanying the fleet was a flotilla of fifty small craft, comprising our mortar and rocket boats and gunboat barges and yawls. When we reached Gibraltar on 9th August, our Rear-Admiral shifted his flag to the *Impregnable,* the second most powerful ship in the squadron and more fitting for Admiral Exmouth's second-in-command. Even more significant was the fact that we found a Dutch squadron there which solicited and obtained leave to co-operate in our campaign.

Through all this passage the utmost care had been taken to train our crews with intense gunnery drills. Every day, Sunday excepted, they were exercised at the guns and on Tuesdays and Fridays the squadron was cleared for action when each ship fired six broadsides. From these and other preparations the crews were thus kept in constant preparation for the battle. They gained new confidence in themselves daily, developed accuracy and teamwork and were in the highest degree elated.

Admiral Exmouth was a masterly tactician. He knew his men would be faced with massive fortifications and that their targets would be enemy cannon framed in solid masonry. Accuracy was essential and this is where his experience in the American War and at Batavia was vital. In these operations he had proved his ability to organise effective assaults from the sea and to devote more than usual forethought to his strategy. Not only was he a gunnery specialist but several of the ships in his squadron, including the *Leander,* had been on the American station and most of his captains were surely chosen for their gunnery experience.

By 6th August we were off the southern coast of Portugal and Admiral Exmouth drafted his first set of orders and distributed them to all captains. Then on Friday the 9th we passed the glorious and hallowed location where the Battle of Trafalgar had so recently been fought and that afternoon anchored in the Bay of Gibraltar. When we arrived we were nothing more than a random collection of ships; when we left we had become a formidable fighting force with a taut expectancy of impending battle. Every day at Gibraltar a myriad of small boats moved continuously between our ships and the shore. The quantities of shot and powder consumed at practice during the last two weeks were as vital to replace as our food and water. In contrast to this was the flow of materials being taken out of the ships for storage ashore.

Our bombardment of Algiers' coastal defences would be returned by red-hot shot and we knew that the risk of fire to the pitch-impregnated timber of ships heavy with cordage and canvas was immense. Admiral Exmouth therefore gave orders that every ship must clear her 'tween-decks of all combustible items which were not absolutely necessary to sustain fighting capability. As a Surgeon, one of my particular interests was in ensuring that personal sea-chests were stowed in the bottom of the hold and a platform built over them to accommodate our wounded. It was also my duty to weed out any men who were not fit for duty on the *Leander* and put them ashore.

My surgeon's mates were equally diligent in counting our stocks of saws, ligatures and bandages. Together we prepared the cockpit for the wounded and made as many tourniquets and pledgets as possible. It has usually been the custom to distribute as many tourniquets as could be spared about the gun-decks. They would then be used by those who carried the wounded down to the cockpit. I have often found, however, that the indiscriminate use of the tourniquet can be extremely prejudical in gun-shot wounds. This instrument is seldom required in naval actions, excepting in operation, and I have long been in the habit of cautioning my assistants against its use, except where it is obviously required.

The fear of haemorrhage is the greatest dread of the seaman, and whenever they are allowed to have tourniquets in the boats or in the tops, they apply them to all manner of wounds without discrimination or proper knowledge. The patient, fancying himself perfectly safe under this protection, creeps into a corner where he is not discovered till the ligature has caused the most direful effects and the otherwise simple wound is rendered highly dangerous.

Another thing I was particularly concerned with was the sponges and bandages to be used in the cockpit for cleaning and dressing wounds. Unfortunately, lint is generally reckoned too expensive for the cockpit and it is common practice to see the same sponge do duty again and again. My own inclination has always been to hoard a much larger store of sponges and bandages than many other Surgeons and not to exchange them indiscriminately among the wounded. By having a hoard of bandages we were enabled to replace them clean every morning, without having recourse to the hard canvas of filthy bunting. Perhaps I am too fastidious, but if the observable decline in blood-poisoning may be attributed to this fussiness, then I am content to be considered wasteful.

The only pause in these activities which united the entire squadron, even if only for a brief celebration, was on 12th August, the birthday of his Royal Highness the Prince Regent. Our flagship hoisted her royal standard and broke forth with a resounding salute of twenty-one guns, and after this lead our other thirty-four vessels discharged their cannon. When this was over the garrison battery took up the chorus. From every cell in it there came jets of fire and puffs of white sulphurous smoke. The Rock glowed like an enormous pastille half ignited and it was near two hours before all the batteries had done speaking.

Then, the day before we left Gibraltar, a Royal Navy vessel arrived from Algiers to warn us that the enemy had been apprised of our approach and were preparing for battle. This news only increased our zeal. We left Gibraltar on 14th August, after some delay from adverse winds, and on the 26th were within twenty miles of Algiers. As the sultry evening drew on we came into view of Cape Cazzina, the northern cape of Algiers Bay. Here we were given clear signs that our approach was known. Lookouts had obviously been placed high in the hills along the coast and they lit great beacon-fires which blazed through the night and heralded our arrival.

Dawn broke cloudless on the morning of the 27th. Even several miles out to sea, we could study the batteries which glowered black-eyed down on to the port's approaches. The Algerines filled every vantage point and seemed almost to welcome us by raising large silken banners which stirred only slightly in the gentle breeze. As we were to learn later that day, Algiers was crowded with guns, garrisoned by intrepid and practised Arab artillerymen and bulwarked with batteries difficult to enfilade and of tremendous strength.

At noon, every man in the squadron was given a proper meal and immediately thereafter the galley fires went over the side. Then the Admiral's interpreter bravely sailed off to meet the captain of the port. He sailed under a flag of truce to offer letters stating our terms to Omar Pashaw, Dey and Governor of the Warlike City and Kingdom of Algiers. Our first and principal article, of course, was the abolition of Christian slavery. A parting sally made to our emissary by one of the ship's officers was widely reported throughout the fleet. 'If you return with an answer from the Dey,' he said, 'and tell us that he accepts our terms outright so that there will be no battle, then we will be so disappointed as to kill you instead!'

Once our man, Abraham Salame by name, had handed over his letters he was kept waiting in the dreadful heat beyond the appointed time limit, within pistol-shot of the walls and an object of derision to the thousands of onlookers who hurled down insults and abuse. In the face of all this activity and the constant flow of signals between our ships, several foreign vessels in the harbour expediently weighed anchor and sailed offshore. By mid-afternoon, when our interpreter had received no reply from the Dey, Admiral Exmouth ordered us to take up battle stations.

Our preparations below were matched by a constant activity on the decks above. Strong rope nettings were hung over the upper-deck under the masts to catch any wreckage or men falling from above. Hammocks were removed and packed around the bulwarks as some protection from splinters and musket balls. Buckets of water were sent aloft and poured over the sails and the booms, and boats were liberally drenched. More buckets were laid by all the hatchways ready for use; wet sand was sprinkled over every deck to make the planking moist and to give surety to the footing of the men. Wet frieze screens were nailed round the magazine hatches so that no spark could possibly penetrate them and wet blankets were hung round each hatchway. And thus our gun-decks were made ready. I believe that no squadron was ever more stalwartly prepared or more resolutely determined for action.

Ours was the first ship together with Admiral Exmouth's *Queen Charlotte* to reach her station and we placed ourselves on the Admiral's larboard bow, at the very entrance of the harbour and close under the massy walls of Algiers. From our vantage point we could easily see that the shore batteries as well as the formidable walls of the town were alive with troops. A host of the enemy crowded on top of the parapets to watch our advance.

In this centuries-old nest of piracy we expected to find a dark spot of cruelty and wickedness. As seen from the sea, however, Algiers appeared a goodly city, very clean and stainless to the eye, with terraces of white marble sporting mosque domes and lance-like minarets. It was built upon the side of a mountain with flat-roofed houses rising gradually from the very shore in the form of an amphitheatre. The town contained many thousand inhabitants; its walls were about a mile and a half in circumference, fourteen to sixteen feet thick, and there was a fortified breakwater or Mole-head formed by uniting an island with the mainland. For the rest it was a white city, gleaming and spotless beneath the blinding sun. Or so it appeared from the sea. But within it was a festering plague-spot of filth and vice and misery.

I am quite sure that the Corsairs were not aware of what they were about, nor what we meant to do. According to their judgement, we should be terrified by their fortifications and not advance so rapidly and closely to the attack. In proof of this I must also observe that they began to load their guns only after the *Queen Charlotte* and almost all of the squadron had passed their battery. For our part, and largely to explain our later success, it was well that we were afforded a good look at the enemy and his situation before we engaged. Across the water we heard Exmouth's crew give the defenders three ironic cheers and then, in silence, both sides waited. Standing on his quarterdeck, the Admiral even waved his hat at the crowd, urging them to abandon their perilous vantage points. He was determined not to be provoked into offering the first fire.

Algiers' reply to our presence, finally, was delivered by gunfire. At a few minutes before three the Algerines, from their eastern battery, fired the first shot of the battle. It was then that Admiral Exmouth, having seen only the smoke of the gun before the sound reached him, said with great clarity: 'That will do; fire, my fine fellows.' Before his Lordship had even finished speaking, *Queen Charlotte*'s broadside was given; it came three times within five or six minutes. At that, the *Leander* and the other ships all did the same. This first barrage was so terrible that they say more than five hundred of the enemy were killed or wounded by it.

Down into the dark narrow streets, into the pillared fountain-courts, at mosque doors and in palace orange-gardens, our shells rolled and hissed, splitting and splintering and scattering death and destruction as their jagged iron flew about in showers. The Dey of Algiers was thus to be sorely deluded by a false confidence in his means of defence and a total ignorance of our naval character.

He had foolishly permitted us to take up our position without molestation, intending to board us from his flotilla of almost forty gun-boats. This flotilla had been assembled on the hidden and sheltered landward side of the Mole-head and immediately began to advance, rashly proposing to reach the flagship and the *Leander*. They were disguised under the smoke of battle but, once seen, tremendous broadsides, chiefly from the *Leander*, caught them in a storm of grape and canister shot and sent all but four to the bottom. After this, the harbour was filled with people swimming from destruction and the *Leander* then trained her guns on the town's principal street. The battle now raged with great fury, officers and men falling very fast, and masts, yards and rigging cutting in all directions.

Our hours of gunnery practice were soon justified. Broadside followed broadside in swift succession until gradually it settled into a long continuous roar as the better-served guns ran up, fired and reloaded at a faster rate than the less efficient guns. It was now a matter for each individual gun captain to follow the drill as quickly as possible but without letting his men become over-excited. Fear or excitement could blind a man to the need for following each distinct step: neglect to swab out the barrel after each shot and retribution in the form of premature explosion was certain. Even the seamen's wives on board the *Severn* helped their husbands by passing shot and powder. All this time, while the *Queen Charlotte* and the *Leander* were setting their teeth into the defences, our other ships together with the Dutch closed up, anchored and joined in the bombardment.

Before the firing had even begun, my assistants and I had stripped to our shirts and rolled up our sleeves to the shoulders. And none too soon, for only minutes later the fighting commenced and wounded began arriving in the cockpit. Up above, the thundering cannon were leaping and banging on the deck, shaking the *Leander* in her every timber. When the battle raged at its most furious and vessel after vessel of the enemy caught fire and blew up, their explosions shook our ship to its very kelson. Below, in our long narrow cockpit, cramped in by overhead beams, the stench of blood soon smothered the evil smell of our tallow candles.

Here, there is generally one inviolable rule — that a wounded man takes his turn, without favour, were he officer or swabber. I did, however, make one carefully considered exception to that rule: all those requiring amputations were singled out and the operations performed immediately. In support of this exception I believe that amputation which so promptly follows a wound is the only effectual means of saving many from the dreadful perturbation and constitutional shock which invariably follows gun-shot wounds some time later. My decision to give priority to amputations was vindicated by our successful outcome: seventeen such operations were performed upon fifteen patients and only four died.

No language can portray the horrors of the *Leander*'s cockpit for a period of thirteen hours. Sixty-five men were wounded and several killed by the first and second broadsides we took; two poor boys were most dreadfully burnt by a red-hot shot blowing up the cartridge which one of them was carefully guarding. The small space occupied for medical accommodation was instantly crowded to excess; without air, panting for breath, bathed in a most profuse perspiration and unable to stand upright, all these men were to be attended to. 'Water, water,' was the incessant cry. Most fortunately we had provided an abundance and the women supplied it liberally. These were the wives of some of the men aboard and their bravery in assisting us in dressing wounds should not go uncommended.

The numbers of wounded arriving were soon so great that some were bleeding to death before we could work our way round to them. Each man faced his pain and fear differently. Some whimpered and moaned, some screamed and damned and begged to be thrown overboard. Up and down the rows went the chaplain and civilians with weak wine and water or lime juice.

This was no time for consultation or studied diagnosis. The wounds are always self-evident, red and horrible; their extent to be taken in at a glance. After that all too brief encounter I had to decide whether to amputate or try to save a limb, and whether a wound was mortal or worth dressing. For fear of my saw so many of those brave men would beg to be allowed to die, but we gave them a gulp of rum and thrust a leather gag into their mouth. After that it took but a moment to make two cuts and to apply the saw. One assistant would hold the poor fellow's body and another the limb or stump which was to come off.

Among so many acts of bravery and endurance, one in particular must not go unremarked. Captain James Wilson, commanding the *Leander*'s detachment, was dreadfully injured by a ball which smashed both his legs. As we were amputating, he saw one of his men nearby. This was one of his sergeants who had just lost an arm. Even in the extremity of his own suffering, the Captain could spare a thought for another: 'Ah, sergeant, are you here? I am sorry to see you thus. This is a glorious work above, we are not unavenged.' This remarkable officer remained conscious and even cheerful until he died a few minutes later.

Those on deck had their own difficulties. Their greatest, in engaging at anchor like this and in a dead calm, was the smoke of battle. Some of our ships could burn more than a thousand pounds of powder every minute during an action. Without any wind, the resulting smoke accumulated between the decks in a thick fog. Little William Chamberlain was later much quoted as saying: 'Jonah in the whale's belly knew as much about the gale outside as any of us here did of the action.' At this stage it is certain that the commanders of both sides ceased to have any immediate control over the conflict, for visibility was reduced to one ship's cable or less. This was now essentially a gunners' battle and the outcome would depend upon the ability of one side to fire longer than the other and to make its shots the more telling.

Our view was so obscured that one of our officers was only able to discern the effect of our fire by climbing the mizzen rigging. What he saw there heartened us greatly. He discovered that the guns in the upper tier of the Mole-head were mostly dismounted and jammed in the ruins of the parapet. Those that were undamaged seemed to have been deserted. A few of the enemy remained hidden behind some coils of cable and from there kept up a troublesome fire.

There was an exhilaration, almost a frenzy, in the speed and physical violence which the gunners applied to the endless rhythm of the drill: load, run out, fire, swab, reload. Our powder monkeys, one of whom was William Chamberlain, could barely keep pace with the speed at which the gunners were feeding their guns. They formed a human chain, passing up fresh charges from the powder lazarette deep in the bowels of the *Leander*. They shouted for more wadding and, when it did not arrive in sufficient quantity, tore off their clothes and slashed them into pieces to pack the gun barrels. One gunner, who was fighting next to William, stripped off his jacket and rammed it home without emptying his pockets. As the gun fired he shouted: 'Damn my eyes, William, I forget to take out my tobacco-box!'

Almost naked, blackened with powder stains, streaming with sweat from their efforts and the extreme heat of the day, our gunners worked on, hour after hour. It is their custom to strip to the waist and take black handkerchiefs and bind them very tightly round their heads and over their ears. Before the firing began they would have buoyed each other up with plans for the prize-money they would win. Most, hardly less merry and as though the prospect of death was remote, made wills — not in writing, but verbally.

In the heat of the action many of the *Leander*'s guns grew at last so hot that they could not be safely used; others recoiled until the wheels made deep troughs in the deck and there stuck; others broke from their carriages. There was little need for the gunners to be given fresh orders. They paused only to gulp a mixture of water and vinegar scooped from the butts lining the centre of each gun deck. Soon neither the gun captains nor the gunnery officers had any real idea of their fall of shot. Conversation was impossible. Orders were given by gesture and hand signals. Men became deaf. William lost his hearing for several days after the battle. But then he was one of the lucky ones; others never regained it.

Deafness was the least maiming of the injuries suffered on the *Leander*. Instant death when a man was struck by a cannon-ball in the head or trunk was usually seen as a better fate than surviving only to undergo the agony of amputation. Dreadful injuries were caused by the mixture of shot which the Algerines were employing: musket balls, nails, broken glass, spikes and pieces of chain. There was also the maiming caused by splinters of wood gouged out of the *Leander*'s structure by the impact of heavy ball. These wounds were among the cruellest of injuries, causing permanent blindness or making deep punctures which resulted in gas gangrene and a lingering death.

When, after more than an hour of this heavy firing, there was no appearance of submission from the enemy, the Admiral determined to destroy the Algerine ships. Accordingly, the *Leander* was ordered to cease firing and the Admiral's barge boarded the nearest enemy gun-boat. The gallant crew fired her so effectively that she was completely in flames almost before the enemy seemed aware of their daring.

Immediately thereafter a vigorous fire was recommenced on both sides, with our flat-boats throwing rockets and firing carronades to such good effect that some of the enemy's magazines were exploded. At first the Algerines thought that our rockets were a kind of signal shot, but when some of them had burst among their troops and houses they then changed their opinion. They assumed they could put out the fire by dousing with water but this only increased the ardour of the flames. These new rockets had been manufactured to stick to wood and penetrate like oil and were thus like nothing the enemy had confronted before.

The *Leander* trained her fire upon the batteries over the town-gate leading into the Mole-head and silenced its last gun just as artillerymen were in the very act of discharging it. At this, one of the Algerine chiefs leapt upon the ruined parapet and shook his drawn scimitar at us. Our fatally pointed cannon had quickly demolished something which, by its brave defenders at least, had been considered impregnable. As evening drew on we observed the city on fire in several places. The Mole-head and other batteries near us were almost completely demolished and the enemy's ships were engulfed in fire.

In the midst of the fury and confusion, with the *Leander* shaking like a tautened rope from the concussion and with blood flowing like bilge water, there was yet a certain order and human purpose which was maintained. The lieutenants walked to and fro about the batteries, regulating the fire and keeping the men to their guns. The powder-monkeys darted here and there on their errands for powder. The half-dozen men told off for cockpit duty came and went with the wounded or paused at the gun-ports to heave a dead or dying man overboard without ceremony.

The hearty and repeated British cheers sent forth on every occasion when houses went tumbling about the enemy not only animated the men on deck, but those who were most severely wounded re-echoed them. Many of the wounded returned to their stations after being dressed and numerous anecdotes could be related of their devotion to their country. Throughout the passage they had been trained to their guns; now they were seen to take aim and fire as deliberately as if they had been exercising.

On our side, after the two flagships it was the frigates which suffered the most, particularly our *Leander*. Her masts, yards, sails and rigging were so entirely cut to pieces from the Fish Market battery as to prevent us from setting any sail. Nothing but the most singular interposition of Divine Providence could have saved us from total destruction. Without a sail to set and the ship a perfect wreck, the wind suddenly veered round and gave the boats an opportunity of coming to our assistance. With the enemy's burning ships drifting dangerously near to us we ran a hawser to another frigate which towed us clear of a fiery meeting.

Among all these gallant deeds it would be remiss not to mention the courage of the Dey himself, who moved from post to post in an open boat under the hottest fire and always urging his men to desperate efforts. They were seen to drop upon their knees and to kiss the skirts of his robes as he passed through their midst. Their guns were well served and handled, and as men fell around them others pressed bravely forward to take their places. But their time was short.

Even for an experienced Surgeon, the carnage I witnessed on board the *Leander* is difficult to describe. On the middle deck I saw eight men lying between two guns, each in the convulsed attitude in which he had expired. One man had an arm extended, the other close to his breast, both fists clenched in a boxing attitude, while his right leg lay by his side, having been cut in two by a cannon-ball close to the hip. Near him was another poor fellow extended on his belly, his face downwards with his back exposed: between his shoulders was buried an 18-pound shot. Legs, arms, brains and mangled bodies were strewn about in all directions. You could scarcely keep your feet in the slipperiness of the decks wet with blood. Here we were faced with a hideous contrast to the pomp, pride and circumstance of glorious war.

But a still more shocking scene was seeing the men and boys who had been on our main gun-deck when a mortar shell exploded in this confined space. Many were crawling about in the most excruciating agony, near naked and perfectly blind. A single feature of their faces could scarcely be discovered and as they crawled they uttered the most heart-rending shrieks and cries, calling out to everyone to put them overboard and out of their misery. I looked about to see if any familiar faces were missing, but it was difficult to recognise my messmates in the curious group of ferocious-looking *banditti* that surrounded me.

Even this, however, falls short of the scene in the cockpit where I worked with our surgeon's mates. Here you tumbled at every step over the dead and wounded. One poor sapper, who was lying in a corner and had been severely wounded by a grape shot through the thigh, on hearing my voice raised himself up and forgetting his sufferings eagerly enquired whether the company had landed and whether we had succeeded. I answered him calmly and, brave soldier that he was, he did not move a muscle of his face until my saw touched the marrow, when he gave such a shriek as I shall never forget.

At about four o'clock a boat was sent to burn the vessel moored across the entrance of the harbour and thus enable the *Leander* to fire more effectively upon the other ships. Some short time later a signal was made to assemble alongside the flagship. All the launches, gun-boats and rocket-boats were then sent to take up their positions for bombarding the harbour and town. By half past seven most of the Dey's fleet was ablaze. An hour later it was utterly destroyed. When the cannonade began to slacken we were surprised to see the stars and to realise that evening had come.

Now I could walk through the lively picture of hell which our decks presented. It was time to order a supply of vinegar to be heated for sprinkling the ship, to drive away the smell of blood and the overpowering stench of battle. In the midst of this survey, exhausted from the rigours of duty and numbed by the scenes of destruction, I was suddenly jolted into a new apprehension of despair by the sight of a young body. It lay face down among a number of others which were only now being thrown overboard. Next to it another boy lay prostrate with the most piercing cries of anguish for his lost companion. I was seized with a dreadful fear that here I had found the dead body of William Chamberlain.

Any soldier, let alone a Surgeon, must steel himself to the horrors and tragedy of battle. So constant and awful is our employment that failure to inure himself could quickly deprive a man of all reason and faith. But although I counted myself a seasoned officer, the sight of this young and broken body sent a terrible tremor through me. I reached out to stop the men in their sorry task and knelt down.

The back of the child's head had been most dreadfully hurt by shot. One half of me professionally appraised the injuries and recognised that death had been immediate. For my other half, though, this was the final and insupportable anguish of the day. With a broken cry wrenched from my lips I tenderly turned the small body over. But even the disfiguring burns across the face could not confuse me. This was not William!

It was, however, one of his young companions in arms. Frantically, my eyes scoured the deck until someone who saw my distress quickly confirmed that William was quite safe. Through all the frenzy of battle my thoughts had still been much on the welfare and role of the child. I was assured that although he had frequently been out of my sight, he had, nevertheless, been an object of tender concern for others in the crew. I was also told, and it in no way detracts from his valour, that he was so fatigued he even slept most soundly for an hour on the quarterdeck during a lull in the action! What a blessed, if temporary, relief from the horrors of battle he had thus been granted.

The dead child was only slightly older than William and his unfortunate story was soon told. He had been a servant to the officers in the wardroom. Late in the battle, the boy had been standing with his young companion on the after-hatchway gratings on the lower deck. The boys were nearly abreast of a gun and at each roar raised their small voices to join the cheers of the crew. One of the men sent them to the fore magazine. Almost immediately a shrill cry raised fears for their situation. Both boys were prone upon the deck; one a shattered corpse, the other quite remarkably unharmed. Burying his head in his dead companion's bosom, the poor child uttered piercing cries of anguish and it was only with great difficulty that the men could separate him from the body of his comrade. In the midst of the battle, surrounded by death and danger, it was impossible not to be affected by this piteous tale. I thanked God's protecting hand that William had been spared a similar fate.

There were accounts from all sides of daring and bravery but one alone had captured the hearts and imaginations of the entire ship, and this was a story that covered William in glory and gave great éclat to his name. Perhaps the daring which I now heard of came from William's misplaced feeling that his exhausted break was not becoming for a sailor, even for someone as young as a ship's boy. Or perhaps he was just exhilarated at the successful progress of the battle and a renewed spirit of patriotic valour fired his blood. Certain it is that the name of William Chamberlain provoked three hearty cheers throughout the *Leander,* something I am sorry to say I had missed while below in the cockpit.

I was eagerly told that as the battle drew on into mid-afternoon, Captain Chetham had cause to come down the poop ladder when all at once the tail of his cocked hat was carried away by a splinter from the bulwarks of the ship. He took off his hat and in examining its damage raised his eyes aloft. There he found more urgent damage still and exclaimed, 'By God, the Union Jack is shot away! Go aft on the poop and tell the signal man to bring me another.' As is usual, the sailor was carrying another flag in his breast, where he had stowed it at the beginning of the action to be ready for any unlucky accident. He came forward and the Captain demanded, 'Who shall nail the proud Union Jack to the fore-royal-masthead?'

According to later report it was this stentorian challenge that really woke little William Chamberlain, rather than any actual explosion from the guns! He leapt to his feet and wove his way towards the Captain with such an engaging display of youthful ardour that no one could have resisted his request. 'Let me, Sir. I can skim aloft like a Sydney swallow.' No one then thought to ask just where this quaint expression originated but the gist of the boy's bravado was quite apparent. 'You carry the honour of the ship in your hand,' said Captain Chetham, as he passed over the flag.

Without speaking, William began to make the best of his way up the two or three tattered shrouds that were left in the fore-rigging. A powder-monkey was never better named, for in no time at all he was clinging with his feet to the royal-mast and hammering away with a serving mallet. The whole deck could not but admire the cool and determined manner in which he accomplished what he had undertaken. If he thought to recover any lost regard from the men who had seen him sleeping during the battle, then this glorious moment was his triumph. A resounding chorus greeted his accomplishment and three times the name 'Chamberlain' rang out to assault the ears of our enemy.

Till nine o'clock, Admiral Exmouth remained with the *Queen Charlotte* in the hottest of the fire, close beside us. Here he was an obvious target for the Algerine snipers who were still crouched behind the shattered masonry of the Mole-head's battery. His example thereby encouraged every man not to give up until the whole was completed, and displayed such perseverance that all were animated with the same spirit. The fire of our ships against that of a brave and desperate enemy appeared to redouble. Although a stout man of sixty-five and worn with service in every climate, our Admiral shouted orders as active and eager as the youngest middy in the fleet. Although he was wounded in three places, his coat torn to shreds with musket balls and his telescope broken in his hand; although he was exposed to the hottest of the fire, he seemed to bear a charmed life.

After ten o'clock, with the Algerine storehouses, arsenals and fleet being all on fire, the burning frigates drifting in the bay, some ten thousand houses destroyed in the city, about six thousand Moors slain and the lower batteries smashed and pounded into shapeless ruins, Admiral Exmouth passed the signal to the fleet to move out of the line of fire, cut cables and make sail. Nearly all the Algerine guns had been silenced, yet in spite of a land breeze springing up this withdrawal was very slow. Most of the fleet was as battered as we were, with little rigging left. Warping and towing, we gradually struggled out into the bay. The whole scene was lighted up by blazing ships and storehouses, and our retiring fleet made an excellent target for the few guns which were left in a fort on the landward side of the town. The occasional growl of cannon did not altogether cease until long after midnight.

With the battle over, the squadron came to anchor in the middle of the bay around midnight. Then, just before dawn, a violent thunderstorm burst upon us. As if to add to the awful grandeur of the scene, the elements thus began their war as soon as our ships and batteries had ended theirs. For nearly three hours the lightning and thunder were incessant and the rain poured down in torrents. This downpour was doubly welcome for it not only cooled the air but also washed more of the blood from the decks which our vinegar had failed to reach.

Finally, at daylight, we were able to muster the ship's company. I joined the Captain, the Purser and other officers to score off the names of those killed. We would find that we had lost seventeen men, including five officers, and that over one hundred other men and four boys had been wounded, some fatally. Our hull had been very severely mauled and it has since been estimated that we had fired over 3500 rounds of shot. From the quarterdeck, the Captain spoke of how pleased he was with the manly way his men had behaved during the action and particularised some of the instances of coolness and courage that had occurred under his observation.

After much cheering he then brought the Purser forward to call over all our names. A profound silence immediately descended on the deck. The names began to be called and each answer was 'Here!', until one name received no reply. 'He was killed, Sir,' said a voice; 'Mark him off, Sir,' said the Captain. Thus it went on and for every few names called there were answers of 'Killed' or 'Wounded' from those who had fought alongside their less fortunate mates.

Now that the action was over, one of our gunners came out from the magazine and said that, although he was almost seventy years old and had been in more than twenty actions, he'd never known or heard of any engagement that had consumed so great a quantity of powder as this in so short a time. This hardly surprised me as I have never seen men so animated as those on the *Leander*. Later, the Algerines themselves very justly observed: 'Hell had opened its mouth upon us through the English ships.'

While the above details show the prominent share which the *Leander* took in the battle, it may likewise be considered as presenting a general view of the zeal and valour that must have been displayed by all. It may be that Britain has sometimes abused her power by sea — for where is the instance, in the annals of the world, of unlimited power unabused? — but assuredly in her expedition against Algiers she has proved that, with her, power is regarded as the greatest blessing when it is exerted in the cause of justice and humanity.

In the morning the Admiral ordered his steward to present a grand supper to the officers of the flagship and all assembled drank to the health of every brave man in the fleet. At noon he sent the Dey a carefully phrased message offering him peace on the terms proposed the day before. Failing the acceptance of those terms, the Admiral promised to renew operations at his own convenience. The Dey, however, capitulated and fired three guns in token of acceptance.

The next day, as the need for my surgical presence below in the cockpit became less pressing, I was able to spend some time above deck and enjoy the welcome sunshine and fresh air. That Friday, 30th August, will forever remain in my memory for its moving testament to our Navy's great victory. It was then that I witnessed the release of over one thousand Christians from slavery. Together with young William Chamberlain, I stood at the rail watching these poor wretches for whom freedom must have seemed a forlorn hope and our bombardment nothing more than an impossible dream.

All eyes were fixed on the two transports which embarked these former slaves who hailed from so many different countries. William was anxious about their ultimate fate, but I could assure him that all would eventually, and mercifully, be returned to their homelands. 'Can there be any Englishmen among those fellows?' he asked, in sympathy for their plight. I replied: 'Almost certainly, William. These Corsairs respect no national boundaries.'

I informed him that England had known their cruelty since the seventeenth century when they had even dared to extend their depredations to the Isle of Wight and the coast of Cornwall and Devon! In short, anywhere there might be people to enslave and goods to plunder, these swift raiders had extended their hungry grasp. 'Women, children and even, once, most famously, a favourite of our Royal Majesty and his Queen were all subjected to this cruelty.' And thus, as we stood watching the embarkation of the liberated slaves, I told him about Captain Jeffrey Hudson, fondly remembered as 'Lord Minimus', who had been held captive here in the middle part of the seventeenth century for over twenty-five years.

'Hudson was reputedly the smallest dwarf in our Kingdom although, unlike other such examples, he was neither ill-proportioned nor uncomely. He had literally been served up to the Queen under an enormous pie-crust and immediately became a particular favourite at the court of Charles I until his capture on a return voyage from France. Eventually he was ransomed and was returned to a family who had given up all hope of ever seeing him again.'

As soon as I uttered these words, their particular poignancy and application to William's own situation caused us both to raise our eyes again to those liberated slaves who would now be returned home. In celebration of their deliverance they had all gathered on the decks and on the shrouds and the yards and were exclaiming and giving cheers and plaudits to every ship in our fleet. In a voice choked with emotion and no doubt enfeebled by his stalwart manning of our guns, William asked a simple question, one which I vowed then and there to honour: 'Can I also be sent home, Doctor?'

This heartfelt appeal was guaranteed to provoke my most urgent sympathy, strengthened even further as it had been by such a show of spirit and gratitude from the liberated slaves. Although it broke my heart, I could do nothing more than promise William that I would personally intercede with the Admiralty to see him sent home after such honourable service in His Majesty's Navy.

On the Monday following, the Dey signed our treaties and the huge ransom which he had previously been paid was returned. For a time, at least, Algiers had her vicious sting extracted. It is true that, unlike Trafalgar, our Battle of Algiers did not cause the death of a national hero. And to cite an even earlier example, it is also true that, unlike Britain's engagement with the Spanish Armada, our action did not take place within sight of the English shore. Nevertheless, the release of so many Christian slaves from the iron fangs of barbarians was indeed an act worthy of Britain; an act calculated to raise the character of her Navy, high as it already stood, still higher in the estimation of the world. Nor will the triumph at Algiers pass to posterity without the name of Exmouth as the leader of the brave band by whose prowess it was gained.

Before we departed Algiers I had a chance to witness more closely the state of the Mole-head. I was greatly surprised to see the horrible damage sustained by the batteries and this fortification. I could not even now distinguish how it was erected, let alone where the batteries had stood, nor the many fine houses which I had seen the day previous. And I observed, too, that they had not more than four or five guns mounted on their carriages, and that of all the rest, some were dismounted and some buried in the rubbish. Besides this, all the bay was full of the hulks of their navy, smoking in every direction, and the water out and inside of the Mole-head was all black, covered with charcoal and half-burnt pieces of wood. But the most shocking and dreadful sight was the number of dead bodies which were floating on the water.

Morale among our Navy was extraordinarily high and some few days before we sailed it was even further augmented by a Memorandum from Admiral Exmouth, which was read to every ship's company.

> The Commander-in-Chief is happy to inform the fleet of the final termination of their strenuous efforts, by the signature of peace, confirmed under a salute of twenty-one guns, on conditions dictated by his Royal Highness the Prince Regent of England. The Commander-in-Chief takes this opportunity of again returning his public thanks to the admirals, captains, officers, seamen, marines, royal sappers and miners, and the royal rocket corps, for the noble support he has received from them throughout the whole of this arduous service; and he is pleased to direct, that on Sunday next a public thanksgiving be offered up to Almighty God for the signal interposition of his Divine Providence during the conflict.

All our wounded were kept on board the *Leander* until our arrival in England, about six weeks after the action. Our long voyage — we did not sail until 3rd September — was occasioned by the fact that the wind set into the westward a few hours after we left. Our slow progress, however, was offset by a great honour: the *Leander* carried Rear-Admiral Milne as far as Gibraltar with Admiral Exmouth's original dispatches. When he anchored there on 10th September, the Rear-Admiral transferred his flag to the faster *Glasgow*. For myself, there was further pride in our medical record.

It is true that the squadron suffered a higher proportion of casualties than had Lord Nelson at Trafalgar. However, only one of the wounded died on that return voyage; no one died on board from disease and no serious case existed on our arrival. This high state of health, I have no doubt, may in a great degree be attributed to the general state of mental excitement kept up previous to the battle; from the moral certainty of its taking place; the constant preparations for it; and the state of exhilaration resulting from the perfect success of the enterprise. Had the excessive exertions required during this dreadful fight been followed by an unsuccessful issue, I have no doubt that my account of the ship's health would have been widely different.

And as shipboard life started to return to normal I even found time to settle to the task of writing a 'Report of the state of the wounded on the *Leander*' for the Commissioners for Transports. This composition then further encouraged me to expand my experiences into this much longer and more personal Account, which will accompany young William home. Once there, it will disclose his own contribution to this day of national triumph.

We arrived back in Portsmouth in mid-October 1816. Bells pealed to welcome us and the harbour soon became crowded with boats full of anxious enquirers for relatives and friends. The welkin rang with cheers from hundreds of voices on the shore and from other ships and was maintained right up until the *Leander* came to anchor. Even in the tumult of such a homecoming I did not neglect a pressing duty; before another month had passed I wrote to the Admiralty [Earl Bathurst] outlining William Chamberlain's pathetic circumstances and soliciting his Lordship's indulgence towards the boy.

His Majesty's Ship Leander, Portsmouth, Nov. 10th, 1816

My Lord,

An entire stranger addressing your Lordship may be deemed presumptuous, but I trust that my motives, when they are known, will plead my excuse, and I shall not trespass further on your Lordship's time than by informing you that William Chamberlain, a boy of eight years of age, was inticed from Port Jackson in New South Wales by the Captain or Crew of the *Frederick*, South Sea Whaler, which ship was captured on her voyage home in the year 1812 by the *Sans Souci*, French Privateer. The *Sans Souci* was afterwards captured by the *Andromache* Frigate, of which Ship I was the Surgeon, and the boy was found on board of her. In consequence of a wound he received in the removal from the *Sans Souci* to the *Andromache* he was brought more particularly under my notice, and from the destitute and forlorn condition in which I found him placed, it naturally excited the feeling of compassion and humanity. I therefore retained him on board the *Andromache* until she was paid off in July 1814, when finding that the poor boy was again to be thrown upon the wide world without friends or any person who might be inclined to take notice of him, I sent him to Scotland where he remained at School until I brought him into this Ship in January 1816.

The boy has been lately affected by Hip Joint disease, and is consequently rendered unfit for His Majesty's Service, and, as my Establishment in Scotland is broken up, I am under the necessity of applying to Your Lordship to request that you would be pleased to order him a passage to his Native Country where he may be enabled to join his parents, who, I am informed, are Settled at Port Jackson; as I understand there are Several Convict ships on the Eve of departing from the Thames, I have taken the liberty of

Soliciting your Lordship's indulgence towards him by ordering him a passage by the first conveyance. The boy is now about thirteen years of age and has good ability. I have myself taught him to read and write, and I shall take upon myself to equip him with clothing for the Voyage.

I am, &c.,

D. Quarrier, M.D., Surgeon, *Leander*.

Five days later, on 15th November 1816, in a rapid and humane response to my plea, young William was himself given the opportunity to authenticate his startling narrative and to seek a charitable resolution to his state. I was required to be in attendance and appeared as a deponent, swearing to an affidavit which summarised my original presentation of the case to the Right Hon'ble Earl Bathurst. That same day William was himself examined on board the *Leander*.

As this missive is intended to tell his story on his safe return, God willing, to family and friends, I include a transcription so that, for once, his own voice can appropriately and movingly be heard. I admit that this Account of mine has been as dry as a sea-biscuit, but by finishing with William's own words I hope to encourage him to eventually append his own text, both fore and aft, to mine.

Examination of William Chamberlain

Questions put to William Chamberlain on board His Majesty's Ship *Leander* at Portsmouth, the 15th day of November, 1816.

Where were you born?
[A] At Port Jackson in N.S.W.

Is your father alive?
[A] He was, when I came away.

What is his name?
[A] James Chamberlain.

Does he reside at Port Jackson?
[A] Yes, Sir.

What is his trade?
[A] A Bricklayer.

How came you to leave him?
[A] I was taken away by the *Frederick*.

Who was the Captain of the *Frederick*, and did he know of your being on board?
[A] I forget his name, but he knew I was on board. He made me very ill, until we were taken by the Privateer.

Did your Father or Mother know of your going away?
[A] No, Sir.

How long did you remain on board the Privateer before she was captured by *Andromache*?
[A] Nine days.

How old are you?
[A] About thirteen.

How long is it since you left Port Jackson?
[A] I believe it is rather more than four years.

Would you wish to return to your Father, if we can obtain a Passage for you?
[A] Yes, Sir.

William Chamberlain

Witness: J. McDowall, Lieut. R.N.; R. C. Holland, Lieut. R.Ms.

A PRAYER IN BATTLE

For many months after my return home in 1817 the family, and soon neighbours and then later even complete strangers, provided engrossed and enquiring audiences for a constant retelling of my life story. I very soon felt like father's old friend John Wilkins, who had so frequently been called upon to take a drink and repeat his account of that day he sailed into history on a whale's back.

I made much use of Dr Quarrier's Account once it was returned to me from Government House, and many was the time I silently thanked the good Doctor for taking the time to compile its narrative. It was easier to give my audience his papers to read than to constantly revisit the past myself. Once only, in private conversation with my father, did I describe the beginning of the battle.

'At first,' I told him, 'a strange noise, such as I had never heard before, arrested my attention. It sounded like the tearing of sails, just over our heads. This was the wind of our enemy's shot. The roaring of cannon was heard from all parts of our trembling ship and, mingling with that coming from our foes, made a most hideous noise. By and by I heard shot strike the sides of the *Leander* until the whole scene became indescribably confused and horrible. It seemed like some tremendous thunderstorm whose deafening roar is attended by incessant streaks of lightning, carrying death in every flash and strewing the ground with the victims of its wrath; only, in our case, the scene was rendered more horrible than that by the presence of torrents of blood which dyed our decks.

'The cries of the wounded rang through all parts of our ship. Those more fortunate who were killed outright were immediately thrown overboard. The wounded were carried to the cockpit as fast as they fell and caused me to wonder how my constant friend and patron Dr Quarrier was faring below. Since I was stationed but a short distance from the main hatchway, I could catch a glance as these wounded were carried below. A glance was all that I could indulge in, for the boys belonging to the gun next to mine were wounded in the early part of the action and I had to spring with all my might to keep three or four cannon supplied with cartridges.

'I saw two of these lads fall nearly together. One of them was struck in the leg by a large shot and then had to suffer amputation above the wound while the other had a grape canister shot sent through his ankle. A stout Yorkshireman lifted him in his arms and hurried him to the cockpit where he had his foot cut off and thus was made lame for life. Two other boys stationed in the quarterdeck were killed. A man who saw one of them taken afterwards told me that his powder caught fire and burnt the flesh almost off his face. In this pitiable condition the agonised boy lifted up both hands, as if imploring relief, when a passing shot instantly cut him in two.'

Apart from this private disclosure, my public retelling of events soon tapered off for I could see that these gory details greatly distressed my parents. Finally, all I would recount was one single, telling fact: throughout the entire battle, and even though I could not claim any particular knowledge of religious truth, I told them that I survived by repeating to myself, over and over again, the Lord's Prayer.

After five years away and having experienced what was often a lonely, cruel and dangerous life, it was a novelty to find myself surrounded by family and friends, all concerned for my well-being. At first I felt singularly honoured. Eventually, however, I began to feel that all this questioning, and even the special treatment I was receiving, prevented the one thing I most desired — finding the life I had left behind in 1811.

Increasingly, I took to long and solitary walks which much alarmed my parents; they never wanted to see me far from their side again. Years later, having young sons of my own, I saw the same dilemma from the other side. Too late I sympathised with the haunted look that would come into my mother's eyes when I embarked on one of my lonely rambles. With hard-won wisdom I could only later understand that these long walks were a fruitless journey seeking to overtake the childhood I had for ever lost.

Sydney-town, my own home, my own family and, of course, most significantly I myself, were so changed that I began to wonder if I ever would be a normal family member again, rather than a resurrected hero of sorts. I can honestly admit that this was a lonely role. I was a stranger to all of my former childhood friends and acquaintances; someone to be pointed out in the street. Now, forty years on, it is obvious that I had seen and experienced more than any thirteen year old should have done this early in life. I wondered if I would ever become more than a cossetted hero, whether I could resign myself to life in a remote colony and an uneventful daily existence.

Without much direction to my days and overcome as I was by the emotion of my homecoming, assistance was to arrive from an unexpected quarter. When I had left my papers at Government House immediately after my arrival home, I imagined that their request for protection and assistance from the Governor was nothing more than a mere formality, something couched in the polite and expected mode of correspondence between two gentlemen and hardly likely to engage any genuine interest in my situation. In this opinion I was entirely wrong and my error came from undervaluing the fervour of Dr Quarrier's Account, which obviously touched Governor Macquarie deeply.

Within three days of my arrival home I was requested to present myself before the Governor himself. My first thought was that Dr Quarrier would not be there to guide me through so great an honour. Immediately I was smitten by a sense of guilt at this unintentional act of betrayal. After all, I had my own father here whose unrestrained emotion on my return was hardly counterfeit.

The Governor's reaction, now that he had read Dr Quarrier's Account, was to suggest that I might compile a similar document of my event-filled months from October 1811 to late September 1812; or, as he expressed it, of the unprecedented events which took a young colonist from Sydney Cove to the Spanish Main. In this undertaking he even most generously offered to provide a well-educated young man who was part of his own household to tutor me for two days a week. This kind offer was most opportune, for the truth was that while I had been educated in the ways of the world far beyond my years, my academic learning was still incomplete. While the good Dr Quarrier had personally encouraged and even supervised my reading and writing, I was lacking in an overall education. And one of the expressed aims of my patient tutoring was to provide the encouragement and assistance necessary to commit my extraordinary story to paper.

My tutor, John Campbell, proved to be a patient and considerate choice. Our first meetings were spent quietly enough as we became acquainted with each other, eventually settling into a daily routine. I would dictate an episode in my life story each morning which he would copy down. The next day he would return with a much improved text which we would peruse together, making corrections and even adding more material which my earnest concentration and his questions gradually elicited. From the very beginning I realised that Mr Campbell would be much more than a scribe. Not only was he my voice to the Governor himself but his historical knowledge and insight added new dimensions to my bald facts and childish interpretations.

Usually he left me to tell my story without interruption. Significantly, one of his earliest additions came in response to the first recollections of my childhood which I shared with him. As a preamble to our written Narrative I recounted the family story of John Wilkins which had so entertained me as a child. When I finished with John riding the whale's back, my tutor sat expectantly as if awaiting another instalment. But this was the end of my version of the events and, when I failed to continue, Mr Campbell was able to take up the account. And thus I found that I had here a collaborator as much as a tutor.

'But William,' he encouraged, 'surely you have heard the sequel to your story of the whale in 1790?'

I confessed that I knew no more, that John's story had always stopped here with him making it safely and heroically to shore. 'There is no more to tell,' I asserted. 'I am certain that I haven't forgotten a single fact in the retelling. It is almost as if I can still hear John's own voice, which rose with excitement as he neared his ending.'

Nodding his head in agreement, Mr Campbell encouraged me but added his own pertinent evaluation: 'John Wilkins was the hero of his own retelling and thus finished with his miraculous escape, but there is more and you shall hear it now.'

FEASTING ON THE WHALE

'The whale seemed to regard Sydney Cove as his own particular domain and continued to range through the harbour and was shortly to be the cause of a very melancholy event late in 1790. At length he ran himself ashore in Manly Cove where he soon attracted whole tribes of Natives. They eagerly flocked there from all parts to regale upon such a welcome and unexpected feast. One of their number, the Native known as Bennelong who had been much favoured with indulgences from the Governor, sent a present of this supposed delicacy to his former patron. Since he had not been seen in the settlement for upwards of five months, Governor Phillip was much gratified by this renewal of affection and resolved to reciprocate.'

Here Mr Campbell paused to ask a question. 'You know, William, don't you, that Governor Macquarie is the fifth Governor of this colony?' I nodded, even though I would have been hard pressed to name the other four. Mr Campbell continued his story.

'Since he was ever ready to promote a felicitous exchange between the two races, Governor Phillip immediately secured articles that he thought would be acceptable as presents. Then he set out for the opposite side of the harbour where the Natives had been sighted assembled round a fire and feasting on the stranded whale. Several Natives appeared on the beach as the boat rowed into the bay, but on its nearer approach they retired among the trees. The Governor stood up in the boat and asked in the Native tongue whether his friend Bennelong was present.

'Having instructed officers to stay by the boat and to have their muskets ready, Phillip then stepped ashore and advanced up the beach with his hands and arms openly displayed. It had ever been his opinion, and what he had observed of these people confirmed it, that the best means of obtaining the confidence of a Native was by example. The Natives, however, did not seem much inclined to approach. Nevertheless, the Governor persevered and "had a parley" with the gathering. At first His Excellency trusted himself alone with them and, after he had shaken hands, returned to the boat and rejoined his anxious party. But once there he ordered a man to return with him, together with Lieutenant Waterhouse, carrying wine, beef, bread and other presents.

'After calling repeatedly on his old acqaintance Bennelong by all his various names, the Governor was finally answered by a Native who appeared with several others at a distance. As the three men moved further from the boat this Native approached nearer. Only when the presents were laid down at a distance of a few paces did he move forward and accept them. Still he would not come near the Governor although, in answer to the question "Where is Bennelong?", he repeatedly said he was the man.

'This, however, could not be believed as he was much altered and his former friend could no longer recognise him at all. In his five months apart from the settlement and the Governor's house in which he resided, Bennelong had grown emaciated and hid behind a wild beard. When the unrecognisable favourite stepped forward, Phillip introduced himself as the Governor, or *beanga*, meaning father. This was a name by which Bennelong had fondly addressed

His Excellency when residing at Sydney Cove. Then, on eight or ten of the Natives placing themselves in a situation to prevent Bennelong being carried off, had it been attempted, the Native came up and held out his hand to the Governor. He was, however, so changed and appeared so poor and miserable that even then there was a doubt whether he were the man.

'At length a bottle of wine was held up and, on the Native being asked what it was, in his own language he answered: "The king." Since Bennelong had always heard His Majesty's health drunk in the first glass after dinner at the Governor's table and had been made to repeat the word before he took his own, he had supposed that the liquor was named "the king". This convinced the Governor that it could be no other than Bennelong himself and every method was tried to entice him to come near, but he always retired on their approach so that they were presently out of sight of the boat.

'After some conversation the Governor returned to his boat and, together with Captain Collins, the Judge Advocate, advanced once more in quest of Bennelong. The Native, who had always been much attached to Judge Collins, now showed himself and with great warmth testified to his satisfaction at seeing him again. He was even heard to call him *babunna*, or brother. Meanwhile, Phillip's men remained on their oars to be ready in case of any accident — a precaution well justified as events soon proved.

'The Natives kept the position they had taken, which showed they were under some apprehension, and the Governor was afraid of alarming them. A promise was extracted by Bennelong that the Governor would return in two days with more presents, most particularly hatchets or tomahawks which were greatly desired. Bennelong also requested the clothes which he had been accustomed to wear but had since discarded. He pointed to a small fire and said that he would sleep there the two nights until the Governor's return. He seemed glad to see his former acquaintances and made himself very merry in recalling the manner of his abrupt departure from the settlement some months earlier. He lay his head on his hand, shut his eyes and said "Governor *nangorar* [asleep]", then imitated the manner in which he had run off in the night. Humour and goodwill now seemed to be reviving the amiability of former times.

'By this stage the cove was full of Natives lured by both the whale feast and the prospect of presents. However, on its being observed that twenty or thirty Natives were drawing themselves into a circle around the visitors, the Governor was apprehensive enough to propose returning to the boat by degrees.

'But it was at this delicate moment that Bennelong, who had presented several Natives by name to his former friends, pointed out another one, whom the Governor now stepped forward to meet. On the Native's making signs that he should not approach further and exhibiting much fear, Phillip threw down his sword. Then he continued to advance, at the same time opening his hands to show that he had no weapons.

'In the course of this interview they now stopped near a spear which was lying on the grass and which Bennelong took up. It was longer than was common and appeared to be a very curious one, being barbed and pointed with hard wood. This excited the Governor's curiosity. He asked Bennelong for it, but instead of complying with this request Bennelong took it to where the stranger was standing and threw it down. Then he took a common short spear and presented that and a club to the Governor, which gave reason to suppose that the spear which had been asked for did not belong to him. Phillip continued to advance.

'The savage, not understanding this civility and perhaps thinking that Phillip was going to seize him as a prisoner or to harm him, lifted his vicious spear from the grass with his foot. Then, fixing it quickly on his throwing-stick, he stood on his defence. But as there was no reason to suppose that he would throw it without provocation, and since he was so near those with whom our party were on such friendly terms, the Governor made a sign for him to lay it down and continued to approach. At the same time he repeated the words *weree weree,* which the Natives use when they wish anything not to be done that displeases them.

'Notwithstanding this, the Native stepped back with his right leg and in an instant darted his killing spear at the Governor. It entered a little above the collarbone and had been discharged with such force that the barb of it came through on the other side. Several other spears were thrown, but happily no further mischief was effected. Immediately after throwing his spear the Native ran off, as did Bennelong and his companions.

'It is certain that this outrage would not have happened had the precaution of taking even a single musket on shore been attended to. The Governor had always placed too great a confidence in these people; he had now, however, been taught a lesson which, it might be presumed, he would never forget.

'Lieutenant Waterhouse, who was close by the Governor at the time, supposed that the attack must be fatal, for Phillip was on the ground and the spear appeared to be much lower down his chest than it really was. He immediately concluded that Phillip had been killed and that there was not a chance for any of them to escape. Turning to run for the boat, he saw Captain Collins fleeing in that direction and calling to the men to bring up the muskets. The Governor himself, having managed to get to his feet, now attempted to run down to the beach holding the spear with both hands in order to keep its end off the ground. But the spear was twelve feet long and the end took the ground and cruelly stopped him.

'On reaching Waterhouse he was heard piteously to plead: "For God's sake, have the spear out." Waterhouse determined on breaking it off and bent it down for that purpose, but owing to its length he could not succeed. Just then another spear came and grazed the skin between the thumb and forefinger of his right hand. He would later explain that this attack added power to his exertions, saying: "With the next sudden jerk I broke the spear off short."

'Spears now began to fly thickly. One of these fell close to the feet of Captain Collins as he was talking to the boat's crew. The sailors, who had by then taken up a position between the wounded Governor and the Natives, fired one of their muskets. This enabled Phillip and Waterhouse to clamber down to the beach where the former, who was by now very faint, was at once lifted into the boat. Collins, with the boat's crew acting as rearguard, then joined the Governor. Immediately they put off and within two hours were back in Sydney Cove.

'At first the Governor would not allow the spearhead to be extracted until he had completed an urgent memorandum. His alarm lest the consequence of the extraction prove fatal meant that Phillip the administrator took brave precedence over Phillip the suffering patient.

'Only then was Mr Balmain, the Assistant Surgeon, permitted to examine the wound; after which he made everybody happy by assuring them that he did not apprehend any fatal consequence from it. He extracted the point of the spear and dressed the injury, and in six weeks the Governor was perfectly recovered. Bennelong's solicitude for Phillip's safety was sincerely shown when every day he came to enquire after the patient's health. It is surely significant, after so serious an affray, that Governor Phillip's sufferings did not prevent him issuing the most express orders against any attempt at retaliation.

'It may naturally be supposed that many would be desirous of punishing what was generally deemed an act of treachery. Governor Phillip, however, did not see the transaction in that light and sternly gave the necessary directions to prevent any of the Natives being fired upon. It was his intention, as soon as he should be able to go out, to have the man given up who had wounded him, or at least some of his tribe. This was not planned with a view of inflicting any punishment but of detaining one or more of these people till they understood each other's language.'

Mr Campbell's account ended here. His final opinion on the near-tragedy was in line with our former Governor's. 'Though Bennelong was glad that Governor Phillip was not killed,' he surmised, 'yet there is no doubt that the Natives throw their spears and take lives in their quarrels, which are very frequent. In this respect they are as ready as the lower class of people in England to strip and box, and think as little of the consequences.

'And this whale, William, that occasioned such trouble,' concluded Mr Campbell, 'was the same spermaceti whose exploits were entertainingly recounted to you over the years by John Wilkins. Now you know the creature's further sad history. Thus might this whale, so oddly linked in the chain of human events, have been the not very remote cause of the death of the Governor after having been the immediate one of that of three other persons.'

Mr Campbell's final verdict on this event was unexpected. 'The wounding of the Governor had one definite effect on our relationship with the Natives. Hitherto, the latter as a body had entirely avoided visiting the encampment. Now they began to come in numbers: men, women and children, and usually without weapons. Within a few months, indeed, a great many took up their abode among us. Bennelong, in fact, was welcomed back into the settlement with such gusto that he begged the Governor to build him a permanent dwelling. A brick hut with tiled roof and chimney was erected at the extremity of the eastern point of Sydney Cove, which has ever since been known as Bennelong Point.'

And here Mr Campbell stopped and cast me a questioning look. I was not too young to realise that this additional narrative extended my own onto a broader stage of events and consequences. Over the following weeks this was to be exactly the role he followed in unravelling my own story and giving it coherence and context. And much of this saga would never have been compiled at all without recourse to a brief journal kept by the *Frederick*'s mate William Bodie, a journal which remained with me after his sad massacre in September 1812.

I cannot therefore claim that the finished Narrative is my own work. The polished prose and the historical insights it displays are testimony to another's skill, someone who was far beyond me in age and education. If I had attempted to tell my story in my own words it would appear as a crinkum-crankum kind of work, comprised as it is of so many twists and turns. What I can lay claim to are its sentiments and the personal adventures which were truly mine, even if they were then filtered through the mature sympathy of John Campbell and couched in his own style. At the very least, his more elegant turn of phrase ensured that my prelude to Dr Quarrier's Account was fit reading for the Governor.

WILLIAM CHAMBERLAIN

BY JOHN CAMPBELL

A True Narrative of a Year in the Life of a Young Colonist.
Compiled by John Campbell for the Hon'ble
Lachlan Macquarie, Governor, with reference to
the Journal of William Bodie.

Look back six years to 1811. It is late spring in our remote convict colony of New South Wales and the shores of Sydney Cove are bathed in warm sunshine. William Chamberlain, a boy not yet eight years old, is gazing across the awesome expanse of water. He himself is not a convict but the son of free settlers. Earlier in the same year a census of the 'Total number of Souls in the Colony' noted that Sydney contained 4895 people — by British standards the outpost had become the equivalent of a small provincial town. On this day in October 1811 William is admiring a new arrival to the harbour, the *Frederick*. Unbeknown to him, an extraordinary fate has arrived on this ship, a fate which will snatch him from his isolated colonial home and catapult him into the very mainstream of European history.

Any difficulty we might have in visualising Sydney Cove as it appeared in 1811 has been conveniently lessened by recourse to a celebrated watercolour graciously made known to us by its owner, the Hon'ble Lachlan Macquarie. *A View of the Town of Sydney,* 1811 by John Lewin is specifically dated 23rd September 1811, and was thus painted within days of William's abduction from Sydney Cove. For all the picturesque conventions which Lewin used — a high vantage point and a vision of untamed land in the foreground acting as a prelude to a distant vista of settlement and human activity — this is still the view, in reverse, which our young sentinel, and the hero of our Narrative, would have seen on that fateful day.

As we look through the proscenium of vegetation our eyes are drawn to the focal point of the whole composition — Sydney Cove itself. There is even a ship of some sort placed stage-centre for us to imagine, still further, that we can lock eyes across the years with our young sentinel.

John Lewin had arrived in this Colony from England in 1800 and for the next eight years devoted himself to natural history painting. Although his 'magnum opus', the *Birds of New Holland,* was published in London in 1808, the connoisseurs' taste for such plate books was already waning. In its place had come an increased market for landscapes and topographic views and, sensitive to the market, Lewin began producing his first known views of Sydney-town. By 1811, as the settlement grew, it was necessary for him to stand back and find vantage points from which his eye could encompass the lateral spread of the town.

The harbour itself was an alluring subject for any artist. But equally tempting was the romantic promise of adventure on the high seas, which could capture any boy's imagination. Just over twenty years earlier, the First Fleet had sailed into this very cove — eleven ships carried 1500 people, of whom two-thirds were convicts. Before 1788 no European had set eyes on our remarkable harbour; Captain James Cook had sailed past in 1770 quite oblivious to its presence. When John White, Surgeon on the Fleet, was faced with its majesty he was eloquent in his praise and called it 'the finest and most extensive harbour in the Universe'.

But the *Frederick* was no convict transport. Almost a generation has passed since 1788 and in that time the infant settlement has grown into a new child of empire. Arthur Phillip, Captain of the Fleet and founding Governor of this colony, had good reason to be proud of his achievement. He prided himself that, although it was a settlement fixing itself upon newly discovered or savage coast, order and useful arrangement had been created out of tumult and confusion.

For some time the settlement waged a desperate battle to become self-sufficient. When supplies from England were lost at sea, and the local harvests failed, it is true that we faced starvation. It is also true that we were daunted by an alien and, initially, impenetrable interior. It was therefore natural that we turned our eyes east to another frontier: the Pacific Ocean and the riches of maritime enterprises. And that is exactly what the *Frederick* sought.

In our quest to illuminate William Chamberlain's unprecedented saga we need a light of pure unwavering intensity. Let us turn therefore to an Argand lamp and enjoy the benefit of its unique illumination. It is fuelled by sperm whale oil and gives about ten times the light of earlier lamps of the same size which had been fuelled by animal or vegetable oil. And it is this whale oil which brought the *Frederick* into Sydney Cove, an English whaler arriving at the beginning of our present Governor's august term of office.

Over one hundred years earlier, in 1699, our English buccaneer William Dampier, then sailing as a Navy captain, had reported astounding numbers of whales in the Pacific oceans: 'The sea is so plentifully stocked with the largest whales that I ever saw.' Yet even with this enthusiastic report, it was not until the 1790s that whale fishing in the Pacifics became a reality. During the voyage of the First Fleet, Captain Phillip reported seeing a 'vast number of very large whales' on the passage out. And once his Fleet had safely arrived in Sydney Cove, a close encounter with a whale in the very harbour itself soon added further credence to his observations.

Then, in 1791, the Third Fleet arrived. One of their transports reported a remarkable sighting of whales when only fifteen leagues out from Sydney Cove: 'We sailed through different shoals of them from twelve o'clock in the day until after sunset. They were all around the horizon as far as we could see from the masthead.' The Captain could draw only one conclusion from this oceanic bounty: 'I saw a very great prospect of establishing a fishery upon this coast.' Opinions like these encouraged the establishment of the whale-fishery that we now enjoy and it was the quest for sperm whale oil which originally bought the *Frederick* sailing into Sydney Cove.

On 9th October 1811 the whaleship was refitted and sailed out of the harbour under Captain Alexander Bodie. The ship left without its full complement of crew but with the unforeseen and last-minute addition of its youngest seaman, William Chamberlain. Once embarked, William had neither the learning nor even the access to pen and paper whereby he could record his unfolding saga. The present Narrative is thus an attempt to rectify that omission and draws upon the voice of the Captain's brother, William Bodie, as heard in his shipboard diary...

8TH OCTOBER 1811

Fully one year ago, it was my melancholy duty to commit my poor brother Alexander into the hands of the Colony's medical authorities. His erratic behaviour had seriously alarmed his crew while on the high seas from England to Sydney-town and had even endangered their lives and the safety of the vessel. On that fateful and lonely voyage, however, no one knew of the troubled seas my brother had crossed and the personal tragedy that had overwhelmed his reason. Since then, I am much relieved to find that a quiet regimen, a gentle and uneventful existence and kind attendance by the senior medical men here have worked a singular amelioration in his condition.

In proof of this, the same medical authorities have now entrusted command of the *Frederick* to its former Captain once again. Additionally, I also sail on the return voyage to England, to oversee Alexander's continued improvement and accompany him with a brother's caring hand.

However, my optimism and my determined good spirits have already been sorely tried even before we embark. In these days before the *Frederick* is due to sail, there has occurred an event so charged with superstitious portent that it has quite unmanned some of the otherwise stalwart crew. To anyone other than a sailor, the roosting of a large black bird atop one mast would be cause for interesting observation rather than dread. But this was a bird which was not seen to alight on any other ship in the harbour, reserving its blessing, or its curse, for the *Frederick* alone. It was never seen to alight in the night and yet for three successive mornings was found brooding aloft, undeterred by the sailors' oaths which sought to dislodge it from its vantage point.

Each morning its departure was only effected when the Captain himself ventured up on deck to quell the commotion. Whether it was his commanding mien or singlar eye, no sooner did he appear and follow his crew's vehement gestures skywards than, with never a sound, the bird spread its funereal wings and soared away. By the third morning the crew were waiting silently on deck for their Captain to expel this malevolent sentinel. Again, it only needed his arrival for the bird to quit its eyrie. No imprecation was needed, or offered; the whole event had now become a dumb show broken only by his gruff order to resume duties.

But this third visitation, and only days before departure, was entirely too much for some of the crew. Wild speculation feeds off the very body of unsettling events as fungi off a smitten tree and in the sailor's experience, more so than the landsman's, strange terrors lurk above as well as below. And of all sailors it is whalemen who, in their hunt, encounter the sea's most appalling visage and challenge it to the death.

A violent argument between some of the more alarmed crew members and a dismissive Captain has resulted in two men and a boy flatly refusing to sail with us. The Captain's problems with his crew have been widely publicised, most particularly when he offered a very substantial reward in the pages of the *Sydney Gazette* for those who deserted. Alexander's anger has been something fearsome. Unfortunately, as he stood on the deck shaking his fist at the deserters and cursing their ignorance, others also began to doubt not only their own safety but the very sanity of the man who would be responsible for their lives during the next year.

His parting insult to these deserters could not easily be forgotten. 'If a ship were bound for Heaven and should stop at Hell first, there are some of you who would still desert her there!' he shouted across the harbour. I pray that the *Frederick* is not hell-bound. I pray that calm seas and even calmer reason will prevail throughout our long voyage home.

My brother is inordinately eager to begin this voyage and his proud words still echo in my ears. 'On land,' he asserted, 'I have the ship's owners and God Almighty to answer to. Here at sea, I *am* God Almighty.' And for all its blasphemy this is no more than most whaling captains would declare. There is no place where a man has greater opportunities of displaying a tyrannical disposition than at sea, where the captain is an absolute monarch and has the freedom to make either a hell or a heaven of the ship. From his decision there is no appeal and to his fiat the seamen under his command must submit.

9th OCTOBER 1811

This day dawned over Sydney Harbour bright and with a good wind, which seemed to offer an auspicious start to our voyage. The hold is packed to over half its capacity with barrels of whale oil and it is expected that the next six months at the fishery will fill the remainder. This bounty has been the profitable work of an interim captain appointed by the government to delegate for my brother during his enforced rest ashore.

Sometime early on this fateful morning, a young boy must have come aboard the *Frederick* quite unobserved by anyone. If his curiosity had been excited by reports of our black sentinel then he was doomed to disappointment. This morning our masthead displayed no such interloper. Even so, it occurs to me as I record this day's sad events, that the missing bird may well have been the harbinger of this other, more tragic and ominous, human visitation. And here I am dutifully bound to set down the absolute truth, as painful as it is, in order to offer extenuating circumstances for the near-death of this child at our hands.

The boy's curiosity soon took him below to the Captain's cabin and stateroom. Suddenly, his foray into the world below decks on the *Frederick* was brought to a frightening conclusion. A loud and completely unexpected sound of breaking glass accompanied by a pitiful kind of howling from the Captain's cabin must have shocked him into a transfixed state. Before he could escape, the cabin door was thrown violently open and a dreadful figure rushed out.

His hands were raised above his head. His fingers spasmed and dripped great gouts of blood so that the boy must have thought he was witness to a murder. I followed behind, beseeching my frenzied brother to stay calm, and grasped his shuddering frame which was racked by vicious fits. Gradually my calm words and soothing voice had their effect. 'Brother, brother. Remember me here. Remember you are on your ship. You are its Captain and all on board are your crew and responsibility. You raise the very Devil when you succumb to these fits. In their madness you smash glass and injure yourself so cruelly. I beg you to listen to my calm reason. We need a Captain's hand and a gentle touch to steer us safely home. We're going home, remember. Home!'

These pleas calmed Alexander somewhat and I implored him, 'Already two men and one of the boys have deserted the *Frederick* and we cannot sail if any more leave us.' All this time his wordless groans had been diminishing until he allowed himself to be settled into a chair. 'Now, brother, I will go and get you something to drink. Steer a quieter course and think of home.'

I placed a hand on my brother's shoulder and then lowered his lantern out of reach onto the floor near where the boy must have been hidden from us in the shadows. A rainbow-coloured halo around the flame showed how bad the air was down below and in this preternatural light Alexander slumped into a chair. He was seemingly out of his frenzy, although an occasional involuntary convulsion of his arm showed that his passions were still smouldering.

It was now, presumably, that the child imagined he had the best chance to make his escape, but some slight noise he made in moving from the corner roused the Captain from his stupor. With a startled cry he leapt to his feet and pointed a shaking hand in the boy's direction. 'Get away from me, you devil. I sail on the next tide and you will not be part of this crew.' I followed the piercing stare and risked a sidelong look at the cabin wall behind him. Alexander was addressing a misshapen shadow cast so monstrously upon the wall. Who was it that had crawled in front of my lantern and was now too frightened to move further?

A piteous and terrified answer came from the dark corner. 'I'm no devil. I'm William, only William. Let me go.' But the way was barred and escape was impossible. Alexander picked up plates and knives off the table and began throwing them at the shadow and now I truly feared for this boy's life. The child stood upright, against all odds determined to flee this scene of havoc. But his sudden appearance in the flickering light only alarmed Alexander still further. With a desperate cry of fear, and energised by the apparent threat of his demon advancing on him, he brought his chair crashing down upon the boy's head.

My first anxiety was that a murder had been committed and I foresaw another dreadful incarceration for my brother, or worse. However, when I realised that William was unconscious and not dead I tended to his injured head. My next duty was to calm my brother, who had realised that his besieging devil was a poor innocent boy. And here I was faced with a dreadful dilemma. We had no idea who this boy was and Alexander truly showed a pitiful remorse for his momentary act of madness. I resolved to take to sea immediately. Once there, I believed, a return to shipboard routine would greatly benefit my brother's peace of mind. And fortunately William soon recovered from the blow to his head, or at least recovered his physical well-being. It was his mind that I could not doctor.

From his own terrified testimony we knew that his name was William but we had no further knowledge of his station or situation. After a dreadful shock and the blow to the head, his own history had become all foreign to him. When he did finally emerge on deck, such was the solicitous regard shown by us towards him that the crew assumed he was the Captain's son. And I must admit that we did nothing to disabuse them of this opinion.

OCTOBER–MARCH 1812

For the next six months we sailed off the east coast of Australia, making a highly profitable swathe through the whale population. By early April we arrived at Norfolk Island and were much delighted with the luxuriant appearance of this little spot. The whole face of the country was dotted with a deep green verdure presenting a scene of the most exuberant fertility. This island had been colonised by Governor Phillip shortly after Sydney Cove and was now a thriving settlement. The number committed to the Governor's charge in 1788 had been greater than could be supported by the inhospitable land, failing crops and the untimely loss of a supply ship. He therefore divided the refractory body into two parts and sent the smaller and more profligate division to Norfolk Island.

It has henceforth been adopted as the ordinary practice that the more abandoned of the convicts and such as have fallen under the sentence of the law a second time should be transported to this island, whose narrow compass and remote location causes it to be much dreaded as a site of punishment.

APRIL 1812

By mid-April our *Frederick* was laden with over one thousand barrels of sperm oil and ready to begin its long voyage home to England. On 11th April we left Norfolk Island and took the sometimes treacherous route back through Torres Strait. In preparation for the boisterous weather we would encounter in the Strait, or certainly beyond, our whaling activity now ceased entirely.

Hitherto our mastheads had been regularly manned and our boats kept in readiness for immediate service, but as the idea of adding any further to our cargo was now relinquished, the lookout for whales was discontinued, the platform on the deck and the fenders on the side of the ship were removed, and the heavy masonry of the try-works for boiling down blubber was broken up and thrown overboard. In short, the routine duties of the crew and the appearance of the *Frederick* were now reduced to those of an ordinary merchantman bearing a cargo to her destined port. And a different and relaxed spirit came over the crew, which even Captain Bodie's alarming outbursts did little to quell.

This welcome respite was not even compromised by the dangers that attended us as we began our westerly passage towards the Indian Ocean. Torres Strait throws up sandbanks at one time which can disappear at another; hence the passage, from being always changed, is always dangerous. It is often the custom therefore, when possible, for ships to go in company, so that if an accident happens to one then the others may lend their assistance. But Captain Bodie would have none of this. His profound relief at being at sea and in command once more would brook no authority other than his own.

Immediately after breakfast each day during our voyage through the Strait, the first mate went to the masthead and, with a speaking trumpet, every few minutes announced our exact situation to the Captain below, who stood on the poop and commanded according to the intelligence received, while the crew watched with the greatest attention to execute his every order.

Our passage lay through reefs and sandbanks which were so near that it appeared as if we could touch them with a harpoon. Wherever we turned there was a different course or channel, but all were in like manner so completely surrounded by those dangerous sandbanks that it was difficult to trace the line; and, if pursued aright, it was so narrow that it might truly be compared to threading the eye of a needle.

For three days we caught bad weather and were forced to scud under bare poles or close-reefed topsails. Then, about three o'clock of the second day, when the wind had abated a little, some of the crew were sent up to shake out a reef from the fore topsail. They had just let go the points when the Captain saw a squall boiling away at the stern. The sea appeared to be carried aloft and he sang out to them to make fast and lay down. They were just below the top when the squall struck and hail pierced us like needles. Before we could say 'Ho!' there were two loud reports: the topsails had been split to ribands. But the worst was yet to come when, in an instant, William Chamberlain lost his balance, fell and ended up overboard.

Although this is every sailor's nightmare, many seamen deliberately choose not to learn to swim. They fear that this skill could deprive them of a swift, merciful end in exchange for a protracted fight with the waves and the horror of a shark attack. William's perilous descent from the masthead was to be an unforeseen blessing – miraculously, its sudden shock restored him to his senses. The abrupt and heart-stopping drama of his fall, the sudden immersion in water and the very real possibility of being drowned were enough to bring back his memory!

From this time on, therefore, the boy's own voice is woven into this Narrative.

'In that instant when I plummeted from the masthead and knew the fear of death, there flashed before me at long last and in this penultimate moment the desperate recall of a most loving presence. In that instant I was restored to the memory of my father. I even imagined that I could hear his anguished warning as I plunged towards the watery abyss. In this way "William, William" rang in my ears and returned me to an identity and a life as I fell headlong. I thus faced a double resurrection as I struggled to surface, both from an overwhelming sea and from waves of recovered emotion.

'Instinctively I fought against the waves until I broached the surface. I saw the huge bulk of the *Frederick* glide beyond me, with pale anxious faces gaping down from the rails. Ropes were thrown but all to no avail and the ship quickly receded. I was left alone struggling in the deep. Eventually the *Frederick* came into the wind's eye and I caught glimpses of its deck as confused figures ran about the poop. They were evidently lowering a boat. Unfortunately, it is true that when there is the slightest swell one cannot see anything from the deck — I mean such a small thing as a man's head — for any real distance.

'The crew later told me that there had been great confusion on board. No one knew that I could swim; no one expected me to be recovered, alive or dead. All had given me up the instant they saw me plunge from the masthead. For my part, the fear which I experienced in the water was beyond description, for all at once my memories of Sydney-town and my family had flooded back and compounded my confusion even further. I was thus overwhelmed both emotionally and physically and in desperation hallooed as loud as I could. Fortunately, someone heard and the rescue boat pulled towards me. They made a drive at me with a boathook, caught me under the shoulder and hauled me aboard.

'A glass of grog was prepared back on the whaleship which I drank with great gusto. However, in case I should feel inclined to go overboard again in expectation of the same reward, I was informed that I should not have it another time. More solemnly, I was made aware that if the man who had been with me on the masthead had fallen, he would have fared very differently as he could not swim. As if to underscore my good fortune still further, the crew regaled me with an account of a former shipmate who had been similarly rescued from the waves. In contrast, this sailor's finite body had been sustained only to permit the infinite one of his reason to drown. When he regained the ship it was found that his ordeal had quite turned his mind!

'But these dire predictions, doleful stories and even my own miraculous escape from drowning were eclipsed by the restoration of much of my memory. It was not surprising that back on the *Frederick* I became the pet of the fo'c's'le for some time afterwards. From this privileged position I even summoned up the courage to question our mercurial Captain and, on this memorable day, learned much about the secrets of navigation. As Captain Bodie was poring over a chart and looking apprehensively towards the horizon, I innocently asked where we were. "Where we should be, William. On board a tight ship under a cloudless sky and carried free before a strong wind."

'After this unsatisfactory reply the Captain returned to his chart and I resumed my vantage point at the rail, where I repeated my question to the Captain's brother. "My boy, the mysteries of navigation are quite foreign to me," he admitted. "Most seamen have no knowledge whatsoever of navigation, particularly the complex process for establishing longitude. You will find that sextants and chronometers are so expensive that sometimes ships like ours don't even carry them."

'My obvious alarm drew forth another admission that was meant to calm my fears. "I believe," he continued, "that it is often the deliberate policy of captains and shipowners to keep their crews in ignorance. If the captain is the only one with the necessary skill and experience to sail the ship safely back to port, then he has a ready barrier to any mutiny." This extraordinary admission made me realise that our erratic Captain held all our lives in thrall to his questionable navigational skills. The rest of the crew, however, hardly seemed to share my concern. The one fact that they were probably more vigilant about than our nautical position was our daily consumption of water.'

'Although my memory had now been considerably restored, there was still much in the past few months which remained lost to my recall, try as I might to bring events to the surface. Indeed, I was never to retrieve anything of my actual leave-taking from Sydney Harbour in October 1811. Its drama was eventually and gently disclosed to me by the Captain's brother, who shared my own name.

'"Captain Bodie," he confided, "suffered a dreadful and debilitating tragedy some years back. Ever since then his nature has been sorely changed and his mind much troubled. In effect, his life has been like a ship's compass which can literally reverse itself when affected by lightning. Then it will point wildly with no certain idea of North. So it is with him."

'This unforeseen revelation astounded me. Although I had no clear memory of my treatment aboard the *Frederick* these past few months, in the last few days I had been lavished with tender care by the Captain. His kindnesses and, I had to admit, the fatherly concern which he displayed towards me had been much in evidence. I shook my head in disbelief but William Bodie resumed his sad account.

'"Two years ago, when my brother arrived in Sydney-town, his erratic behaviour had so alarmed his crew that a medical committee demanded he stay ashore for the next year in the hope that a quiet life and soothing care would restore his reason. My brother had become very impatient of control and jealous of his power and station in the ship. He would often roam the deck at night and alarm the watch by declaring that the Devil was in his cabin. When the time came for the *Frederick* to return to England he resumed his command. Because I was not entirely comfortable with him facing the lonely months at sea again I signed on as first mate. And sadly, I have to admit, William, that my brother's restoration has not been entirely successful."

'Here he looked quizzically at me as if expecting some reaction. "The Captain has been like a second father to me these past few days," I suggested, "although one thing does trouble me. That is, how I came to be on board the *Frederick*. I'm surely too young to have signed on as a crew member, yet it is all lost to my recollection."

'William Bodie took some time to answer and I grew fearful that he would disclose some calamity that had befallen my family and seen me consigned to this life at sea. My stammering questions and pleas in this vein, however, brought a quick denial. "I have no knowledge of the family you left behind, William. Indeed, your father knows nothing of your whereabouts and may even presume that you . . . that you are dead!"

'My confusion and consternation at this news were extreme. In an effort to calm my agitation he continued, "I promise you, William, that as true as the circumference is to its centre, neither my brother nor I meant you any harm." And now William Bodie proceeded to reveal my forgotten history, submerged and unfathomable to my most desperate efforts at recollection.

'"My first anxiety," he admitted, "was that a murder had been committed and I foresaw another dreadful incarceration for my brother, and worse. However, when I realised that you were only unconscious and not dead I tended to your injured head. You must understand that I was faced with a dreadful dilemma. We had no idea who you were and my brother's pitiful remorse for his momentary act of madness strengthened my resolve to take to sea immediately. Once there, I believed, a return to shipboard routine would greatly benefit my brother's peace of mind. Fortunately you soon revived from the blow to your head, or at least recovered your physical well-being. It was your mind that could not be doctored. Your name alone was the only knowledge we had of your being. After the blow to your head, your own history was all foreign to you and so when you finally emerged on deck as a cabin boy you were introduced as William Bodie, my own namesake and . . ."

'Here the fellow stumbled in his account and appeared much overcome by a painful memory. I waited quietly, although I burned with questions, and presently he surfaced from his sad reverie. "After your physical recovery I could chart the beginning of a new equilibrium in my brother's troubled mind and I could not disabuse him of the idea that you were his own child. No one else knew that his real son had been lost at sea over three years previous, a loss which drowned a father's fond hopes and completely reversed the compass of his reason. And so far as the crew know, this is your relationship. Then, when you fell from the masthead, William, he truly reacted as if he had lost his own flesh and blood."

'I was greatly taken aback by the disclosure of my dramatic initiation aboard and found it difficult to reconcile the description of the deranged Captain with the caring figure I knew. Finally, my mind stopped wrestling with unanswered questions and quietened down. My thoughts returned to my rescue and resurrection and its moving effect on both brothers. I could never doubt, whatever might have happened in the past, that they were not deeply affected by my plight and vigilant for my well-being. When I raised my eyes to my sad narrator I clearly saw that here was a stalwart friend in whom I could trust. And in recognition of this he concluded his sorry tale.

'"My brother's only child, about your own age and as lively aloft, was making a much anticipated first voyage with his father. The Captain, from an unfortunate mishap with a cutting lance the day before, did not himself launch his own whaleboat for the chase but entrusted it to another. He also trusted his son to the experienced hands of his crew and, without any apprehension, watched them sail away in pursuit of a whale. It proved to be a more arduous battle than anyone had anticipated and as a louring night sky roiled overhead the Captain began to search the horizon for two of his whaleboats which had not returned.

'"Nowhere could they be sighted. This is not an infrequent occurrence, as a whale can tow a small boat far ahead of the ship, but paternal anxiety grew apace with the violence of the sea until all hands were enjoined in an anxious scanning of the horizon. The masts and yards were all thickly crowded with men on lookout and a fire was kindled in the try-pots as a beacon. At last one boat was sighted but, greatly as its men were welcomed aboard, it was their companion which the captain desired had been found in their stead.

'"For three dreadful weeks he kept his ship cruising, criss-crossing the area where the whaleboat with its precious cargo had last been seen. The frantic circles his ship made in this desperate search have been echoed in the frenzied turmoil of his mind ever since."

'The pathos of this story quite unmanned me, and if it had not been for William Bodie's heartfelt words of comfort and his compassionate embrace, then I know that I should have been in a flood of tears. From this day on I charted a new course. I would not disabuse my shipmates of their notion that my father and uncle were in charge of the *Frederick*, nor would I let them imagine that I was not a willing voyager.'

WILLIAM BODIE'S JOURNAL: TIMOR TO MAURITIUS

JUNE 1812

The Captain intended to make a landfall at the Dutch island of Timor. On its western coast there are several bays and it was one of these, Soutranha, which was to provide the *Frederick* with anchorage and supplies. This was a popular port of call for whaling ships on their return voyages and a large proportion of its commerce was derived from the visits of English South-Seamen who called there during the Timor season to procure refreshments. No pilotage or harbour-dues were demanded, permission to take wood and water was freely bestowed and every essential supply was readily obtainable. Provisions were dear, if purchased with specie, but as the Natives preferred barter, and chiefly required European manufactures such as muskets, ammunition and cutlery in exchange for their commodities, the balance of trade was much in our favour.

But by far the most sought-after items in these transactions were the teeth of the sperm whale. When solid and of large size they were eagerly exchanged with the Natives who carved them into treasured ornaments or into a sheath for their fearsome Native dagger called the *kris*. Surprisingly, for all their interest in obtaining this ivory, the Natives did not ever appear to hunt the whale themselves even though the creatures frequent their abrupt coasts. In contrast, many English South-Seamen cruise in the Strait of Ombay and adjacent seas during the winter months, and usually with success. It is surely remarkable that American whalers, though seldom behind the English in enterprise, have not as yet ventured their barques into these waters.

By early June we had arrived at Timor and the harbour of Soutranha. The Malay village on its shore was hedged with jungle and bordered with mangroves and brackish marshes which we knew were most pernicious to the health. Native dwellings were scattered over the plain in detached groups and embowered by cultivated enclosures and groves of palms and cocoa-nut. These thatched huts were raised by posts to varying heights above the ground and, although they seemed but flimsy structures, their bamboo construction yet gave them a surprising strength.

The population was very unprepossessing. They were short in stature, small-boned and even meagre. The men rendered themselves more unsightly by adopting long and tangled locks of coarse black hair and by eating the betel-nut which stained their mouths an unsightly red. Their singular attire was well suited to the climate and seldom exceeded a piece of cloth worn round the loins and a handkerchief covering the head. Silver or ivory bangles, as well as necklaces composed of coloured beads or of small fragrant black and yellow seeds, were the favourite ornaments worn by both sexes. Even more colourful were the parrots' feathers tied to a strip of bamboo and worn in the ears and bracelets of long white hair which festooned the ankles.

But the most remarkable mode of adornment which we observed there, and that exclusive to the males, was to cover the front teeth with plates of silver! Then the gleam of polished metal, displayed within the parted lips, added to the men's naturally ferocious looks and produced an extraordinary and somewhat diabolical effect. We were informed that this singular decoration was an order of military merit, conferred by the Rajah upon those of his warriors who had distinguished themselves in the field.

We had no sooner set foot on shore than we were surrounded by half-dressed copper-coloured children, covered with cockatoos; that is, they had these birds of all sizes and degrees of beauty perched on their heads, shoulders, arms and hands, or else carried sticks on which the birds roosted. The cockatoo, usually so wild, appeared in their possession as tame as a parrot; but the noise and number of the birds, with the chattering entreaties of their owners for us to purchase, quite bewildered our senses.

And it was here that the Captain suffered another unpleasant altercation with his men. This time he took decisive action and simply abandoned them. Later events would prove the dreadful folly of this rash decision which reduced the *Frederick*'s strength to only nineteen men. His morose temper remained as unpredictable as ever and the crew were worked as hard as if the success of the voyage depended upon their ceaseless toil of scrubbing, scraping and polishing. Discipline was maintained at a high pitch of perfection, no man daring to look awry, much less complain of any hardship, however great. Yet even this humble submissiveness could not satisfy their tyrant and at last his cruelty took a more active shape.

One of the crew must needs try his hand at making a villainous decoction which he called 'beer,' the principal ingredients of which were potatoes and molasses. Since potatoes formed no part of the crew's diet, he set his wits to work to steal sufficient for his purpose and succeeded so far that he obtained over a dozen. The fact that his theft was soon discovered became painfully evident when our companion and his brewage were dragged to the quarterdeck. There, in the presence of all hands, he was arraigned, found guilty of stealing from the ship's stores and had sentence passed upon him.

By means of two small pieces of fishing line he was suspended by his thumbs in the weather rigging, in such a manner that when the *Frederick* was upright his toes touched the deck, but when she rolled his whole weight hung from his thumbs. One would have thought that this of itself was sufficient torture for any offence. In addition, however, he received two dozen lashes with an improvised cat-o'-nine-tails, laid on by the brawny arm of one of our harpooners. We were all compelled to witness this punishment and our feelings may be imagined. After what seemed a terribly long time to me, the poor fellow fainted. He was then released and the crew were permitted to take him forward and revive him. As soon as he was able to stand on his feet he was called on deck again and not allowed to go below till his watch was over.

Meanwhile, the Captain improved the occasion by giving the men a short harangue, the burden of which was that they had now seen a little of what any of them might expect if they played any 'dog's tricks' on him. But you can get used to anything, I suppose, and after the first shock of the atrocity was over, things went on again pretty much as usual.

In mid-June we sailed from Timor with the land-breeze and steered to the S.W. along the coast. Two days later we passed the island of Savu situated at the S.W. extremity of the Strait of Ombay, entered the Indian Ocean and now crossed the ordinary track of outward-bound ships from England to India as we headed for Mauritius.

AUGUST 1812

The *Frederick* sighted jagged peaks at first light early in August. It had been less than two years since British and Indian troops had taken this island from the French and the 'Ile de France' had now reverted to its former Dutch name of Mauritius. After so long a voyage the sight of any portion of land is doubtless delightful to the traveller; but how much more does it appear welcome when he knows that he shall find on it the men, manners and language of his Mother Country.

Land clouds rising in the distance ahead of the *Frederick* told of Mauritius while the island still lay invisible in the vastness of the ocean. We began to take soundings at each change of the watch and before long caught the dull distant rumble of breakers. By daybreak we reached the small islands to the north of Mauritius and by early morning we had doubled its south-east point. Wind and tide drove us southwards until we were running along the coast in clear view of the shore.

The lowlands here were covered with sugar cane plantations and interspersed with cottages, villas and hamlets which stood out against a dramatic backdrop of volcanic eminences. These mountains rose in the misty majesty of the morning, emerging and sinking in their sea of vapours that now encircled, now overflowed, and finally deserted them.

The *Frederick* carried no charts of the island and we were forced to wait until we could follow a local schooner into Port Louis, so that it was not until evening that we finally came to anchor in the harbour. As it was too late for the inspector to visit us and examine our bills of health, we remained on board till the next day. Even though we were ahead of the malignant onslaught of summer, this first night and the others which followed were vexsome. A veritable world of insects seemed to overtake us. Our sweating bodies were the targets of endless stinging armadas, moths blanketed the lanterns and anonymous vermin soon found their way into every corner of the ship.

Even so, after ten weeks confined, our spirits were exhilarated at the prospect of liberty. And yet even this desire was tinged with sadness when we recollected Matthew Flinders' hostile reception here almost a decade ago. Under the new British regime we were naturally assured of a safe landing, although in December 1803 our celebrated countryman had been much less fortunate. It was then that Flinders had landed in search of urgent repairs for his ship, completely unaware that war had again broken out between England and France. By a cruel twist of fate he was to be held captive on this island until June 1810.

But now everything looked peaceful and flourishing. The only 'threat' came from the island's singular and grotesque peaks. These bore names as fanciful as their appearance; one pair was degraded by the appellation of the 'Ass's Ears'. The island was of an irregular oval form, nearly surrounded by coral reefs of red, white and black, with shallow lagoons between them and the shore whereby access is rendered difficult. The sky here was remarkably clear and of an indescribable purity of blue; and the mountains stood out in bold relief. This effect allowed the eye to look beyond their majestic outline to an unfathomable depth of ether.

We had arrived on a Sunday yet the whole town wore the aspect of heathenism on this Lord's day. No shops were closed, except for the convenience of the owners; pleasure and business occupied the free, and drudgery and degradation marked the slaves, as on other days. We understood, however, that even in these respects, society, in outward decorum, had been much ameliorated in the eighteen months since the island had fallen into the hands of the British.

Slavery existed here to such an extent that its miserable victims were met in droves, or singly, everywhere, performing all kinds of base, penal and brute labour. In the streets they were to be seen dragging carts and drays, like beasts of burden. But it is on the sugar plantations that these helots swarm and blacken the face of the country. Such estates are rather numerous than large, and are tilled by bodies of field-drudges numbering in their hundreds. The Negro family here, as in the West Indies, is the nursery for this monstrous injustice. These slaves were principally obtained from the east coast of Africa while the trade was permitted. Now, although it is prohibited, there is reason to believe that it is still managed in the usual evasive manner, no preventive service being sufficient to suppress the contraband of blood in a slave colony so distant from the mother country.

Now that the boy's strength was fully restored and his spirits high, the prospect of a visit ashore to this picturesque island was very tempting. Seeing his eagerness, I offered to take him to the island's Bazaar.

The vibrant colours and the heady perfumes in the tropical air from an astounding display of merchandise were matched by the most extraordinarily diverse population. We were astonished to see that slaves carried wealthy settlers about in palanquins like royalty itself. A wide avenue traversed the whole of the Bazaar and on each side were large covered sheds with a raised pavement on which goods were displayed. The fruit stalls were a cornucopia of tropical produce. Much of it was unfamiliar to me; even bananas — which I had eaten at home — were present in the most exotic forms. There was even one red variety over a foot long! William remarked on this novelty when he discovered a pet monkey expertly peeling back its skin and daintily eating the fruit.

The vegetable stalls were equally engrossing and the noise was fearsome. After weeks at sea it was painful to be assaulted on all sides by its unrelenting volume. The bargaining from the excited and high-pitched voices of Creoles and Indians made it a very Babel and large cages of native birds added their own unique contribution. These included pretty scarlet cardinals with their brown mates and greenish-yellow canaries whose sweet song was completely lost in the mêlée. Rare birds were offered here by sailors from foreign lands, yet none were as bright or as expensive as those from New Guinea. As well as birds, I doubt if any place in the world could have offered such a variety of fish, many of them of such brilliant colours that I can only compare them to the gorgeous plumaged birds we had seen in Timor.

Nearby was a place set aside for dried provisions, sold mostly by Arabs who exposed their wares on the pavement in small sacks or baskets. It seemed that every known species of bean was offered, as well as rice, maize, colourful spices, chillis, coffee and even dried herbs and drugs. The air swam with urgent perfumes: wood smoke and horse dung, of course, but also exotic and novel aromas like saffron and musk. There seemed to be scarcely a leaf or bark which was not used by Creole or Indian for some ailment. They have the most profound faith in herb teas. I was much taken with these and, having found someone to interpret for me, I then tested the vendors' knowledge of drugs which might alleviate my brother's condition.

'These Natives know of clever tisanes, William,' I told my young companion as I gazed in wonder at the colourful display. 'Among these healing herbs there are also many which no doubt contain dire properties. I am sure that they use them freely when prompted by revenge or other passions.'

Most of the vendors spoke a little English of which they were very proud. When they perceived that we were newcomers and somewhat awed by their noise and display, we acquired quite an entourage with all vociferating 'What you want? Come to me.

He no good' and so on, and each offering us the most incongruous articles. Once we were rash enough to show any interest in something, we were deluged with offers.

Ironically, the first things to be offered to us were whale's teeth! It appeared that no jewel, however valuable, was half so much esteemed in Europe or America as was a whale's tooth here. We saw the Natives laugh and cry for joy at the possession of one of these darling treasures. Ivory, however finely wrought, bears no comparison in their estimation and is worn only by the lower classes. Even then, it is made into the form of whale's teeth to be worn as ear ornaments.

In the centre of this Bazaar was erected a stand on which sat a large pair of scales attended by a municipal officer. Anyone being aggrieved by receiving short weight could have the disputed article reweighed and, if found wanting, the seller would be arrested, heavily fined and lose his standing in the market.

Together with my young companion I spent one of the most carefree and entertaining days here that I can ever remember. At gunfire in the early evening, all the Lascars burned a kind of frankincense in their scales and about the stalls, muttering prayers over it to keep away the devils, to bring them good sales the next day and to render their houses lucky. Their heady smoke filled our senses and I prayed that some of its supposed good fortune would waft over us.

THE DEVIL OF THE WATERS

SEPTEMBER 1812

After we left Mauritius and upon nearing the Cape the weather became as turbulent and treacherous as is proverbial to this region. At night, when the wind blew off the shore, the odour of land was very perceptible and a large quantity of fine sand was deposited on our sails. We rounded the Cape without approaching very near to it, but three days elapsed before the winds permitted us to shape a direct course for St Helena. This remote location is situated in the midst of the Ethiopic Ocean and is believed to be the most distant island from a continent of any in the world. Three weeks later we fell into its parallel of latitude about 50 or 60 degrees to the east. Cautious of missing so small a landfall, we decided to lie by all night for fear of running past in the dark.

It was then, on one sad morning, while reefing the fore-topsail, that a man fell from the yard and instantly met his fate on the unforgiving deck below. With understandable haste, the body was taken below where it was neatly sewn up in canvas with a heavy piece of sandstone secured to the feet. It was then brought back on deck, laid on a hatch at the gangway and covered with our flag. Then all hands were mustered. The silence of death hung over us all, broken only by the tolling of the *Frederick*'s bell.

Our Captain stepped forward into this gloomy throng with a prayer book in his hand. True, the man was of no special note among us, but death had conferred upon him a brevet rank by virtue of which he dominated every thought. It seemed strange to me that we, who faced danger so often and so variously that natural fear had become numbed by custom, should now be so impressed by the simple, inevitable fact. I suppose it was because none of us were able to realise the imminence of Death until we saw his handiwork.

Alexander turned the pages of his prayer book to the burial service and began the majestic sentences: 'I am the Resurrection and the Life . . .' At the words, 'We therefore commit his body to the deep', he paused and, the mate making a sign, two of the harpooners tilted the hatch and the remains slid into the unknown depths. Several of the dead man's companions murmured prayers for the repose of his soul, but matters soon resumed their normal course. The tension over, back came the strings of life into their former position, to play the same old tunes and discords once again.

The sudden finality whereby one man could step from life into eternity in an instant cast a pall over us all and it was a sober second dogwatch that night. But this tragedy was not the end of that long day's sad unwinding. Towards night an inexplicable intensity in the air seemed to gather. If I had been a superstitious man I would have called it a summoning of sorts, for the sultry night wore on until the simmering dread on board the *Frederick* was unleashed.

In the middle watch of the night, in the murky damp of the tropics, an unsettling display of St Elmo's fires was seen about our mastheads and the tips of our spars. These pale lights, sometimes as large as a man's head, glowed with a peculiar and preternatural light. They continued so long and burned so steadily that they excited the speculation and superstitions of all the crew.

'Look aloft,' cried the mate on watch. The yardarms were tipped with a pallid fire and each mast was lit like a gigantic wax taper. Unfortunately, although the hour was late the Captain was present on deck, having come up from his cabin at the changing of the watch. Although he had undoubtedly witnessed such displays before, in his present and distracted state of mind the effect was most unwelcome.

'See you! See you! God's burning finger smites our ship,' he exclaimed. For those who had never seen such a phenomenon before, this apocalyptic warning greatly increased their feeling of dread. And when William stumbled up on deck to witness the commotion, his fright was only increased by the words of one of the crew.

'These be corpusants,' he was told, 'the ghosts of men who have died aboard the *Frederick* and now return to warn us of the approach of further death!' Coming so quickly as these apparitions did after the death of one of our crew, and heightened by the dire words from the Captain, a dreadful apprehension swept through the boat. The crew were now prompted to believe that another tragedy was imminent. And I cannot but feel that my brother's already over-heated mind was further inflamed by these grisly and unsettling events.

OCTOBER 1812

For some days thereafter my brother was not seen on deck. It is, of course, usual for the Captain to stand no watch, and because he seldom has any conversation with the crew and issues orders through the mates he can be the most solitary figure aboard. For these reasons they had not been overly concerned at his absence from the deck. However, when they saw the harrowed look on his face they knew that once again he was wrestling with his devils. I had resorted to locking the way aft to the Captain's cabin lest my brother appear and alarm the crew with wayward words and erratic actions. Eventually, though, when distant sails were sighted and the chance of an exchange between ships was probable, the Captain did make an appearance.

Every sailor loves a 'gam', as these meetings at sea are called, and in expectation that this welcome novelty would distract his tormented thoughts, the Captain was helped to the rail and given a speaking-trumpet to carry his voice across the water.

'This is the *Victoria,*' we heard, as our companion hailed us. 'What ship are you?'

The Captain stared at our neighbour without raising the trumpet to his lips. It seemed that he was not going to answer and our crew could be heard apprehensively making their own subdued replies. But then, finally, the old man seemed to gather his wits, although his answer when it came was as unexpected as it was awful.

'This is the Ark of Blood,' he offered in a wild and trembling voice, 'and it is commanded by the Devil of the Waters. Like Lucifer we are fallen from glory and bound for Hell.'

Mercifully, our visitor was deaf to this dreadful introduction, although all on board the *Frederick* were deeply troubled. With much difficulty I calmed my brother and took him below. By the time I returned on deck the *Victoria* was much closer and we learned that this was one of a number of ships that would sail together from St Helena for protection from French privateers. We were very strongly urged to take advantage of their escort.

Soon after that we came in sight of the celebrated island, but found that its appearance was in no way as inviting as Mauritius. Brown and barren rocks rose perpendicularly out of the sea like an imposing array of fortification and the entire coast presented a rampart of weathered and iron-bound cliffs of a sombre burned hue – scarce a trace of vegetation was visible from the water. It was altogether melancholy in the extreme. Towards noon we opened the more lively scenery of James' Valley and cast anchor in St Helena Roads. On landing, the English aspect which everywhere prevailed seemed to unite us once more to civilised life and to the national attractions of our native land. This island is ideally placed to advantage for fleets returning from the eastern seas as a spot for rendezvous and supplies. However, in a political light, the necessity for preserving this speck of land, at no inconsiderable expense, as an important key to our eastern territories should cause the British to regret that it had ever emerged from the ocean.

All accessible parts of the coast are very strongly fortified, but more especially the entrance to James' Valley and the cliffs that command the anchorage. Many of the mountains whose pinnacled heights appear inaccessible are defended by pieces of ordnance, slung in chains or carriaged on their rocky summits. Telegraphs, communicating from hill to hill, convey intelligence over the island with extreme rapidity. On our arrival here we joined the English convoy preparing to sail for home. Although he clearly disliked the idea, the Captain was persuaded, for the moment at least, of sailing in company.

Two nights after the *Frederick* sailed from the island we found ourselves some miles to the leeward of the convoy. Truly, I believe that all aboard this unfortunate vessel are on a hell-bent and solitary course, sailing headlong into melancholy and madness. When our crew could no longer discern the lights of the nearest vessels they began to grow apprehensive. Perhaps this suited my brother's lone temperament best, as he clearly chafed at sailing in convoy and under someone else's command. My earnest petitions for our personal safety, let alone for that of the *Frederick* and our crew, had apparently been cast adrift.

We separated from the convoy off Ascension Island and with ill-considered bravado sailed alone into dangerous waters towards the distant English Channel. In sad summation of this melancholy time I can say only this: when I look into the Captain's eyes, I see on their horizon the certain clouds of gathering madness.

And it is here that William Bodie's journal comes to its abrupt conclusion. The final confrontation and its dreadful climax must now be the product of my own words. This, however, has proved to be no easy task. As William's tutor, and now his scribe, I have tried not to intrude my voice or weave my interpretation too intrusively into his story. I can still see how the young boy is shaken by the memory of this sad finale. In recollecting the *Frederick*'s fatal meeting with the French, he resumed in a halting and tremulous voice.

'Within hours we saw two strange sails bearing down on us and we remained in a state of anxiety for some time, not knowing whether they were English or French. Scarcely had our first emotions of alarm been raised when a particular manoeuvre by the strangers clearly manifested their hostile intentions. By a dreadful fatality, the wind, which had until then blown with some waywardness, now bore us still further away from the convoy and towards these strangers. When spy-glasses could eventually discern the newcomers crossing the bow, it was to announce that they were French privateers, far outweighing us in tonnage and cannon. This was a dire warning to anyone who imagined that the high seas were without peril or that erratic behaviour could be without fatal consequences.

'Our ship now became a scene of great apprehension and confusion; the Captain and mate were so much embarrassed that they did not know what course to pursue for the safety of the vessel. There was no alternative; no possible way of escape could be devised. Those that were of the opinion that a signal gun should be fired to alert the convoy's commodore were soon appalled to learn that he was well beyond hearing range. Others were of the contrary opinion and imagined that any signal, however hopeless, would only serve to enrage the Frenchmen, who would undoubtedly revenge themselves if the *Frederick* fell into their hands.

'The Captain was determined to make his escape, if it were possible. Accordingly, he immediately commanded the crew to shake out all the reefs and loose the top-gallant sails. These sails being all spread, he put up his helm and vainly attempted to steer for the convoy. But the Frenchmen had not been idle during this time. They had recognised these evasive movements and, once they discovered our intentions, they immediately made all possible sail and altered course. The wind suited the French and we could not show our heels to any advantage. Unlike our sturdy whaling boat, their vessels were sleek and fast and soon bore down upon us.

'It was clear that they were determined to pursue and undoubtedly had marked us out for their night's prey; yet, notwithstanding the discouraging position in which we were placed, there still remained a universal hope and persevering spirit among us. But this hope was only momentary and it faded as the privateers drew nearer and nearer. The enemy, having approached us on our weather quarter, hailed us in broken English and commanded us to heave to, with a head to the westward, and that without delay.

'"I will speak to them in a language we all understand," Captain Bodie cried. "I will not part with so fine a ship and cargo without a deadly struggle." And so, although he knew that disobedience to the French command might be attended by the sacrifice of some lives, he refused to take in sail. The French immediately fired a gun, the

Frederick answered, and a sharp action commenced which continued for about a quarter of an hour. While the *Frederick* carried fourteen guns, the Captain's disagreeable relationship with his crew had fatally reduced us to nineteen men. In comparison, the French carried fourteen guns and one hundred and thirty men on one vessel alone.'

William's sad words summon up all the terror of that meeting: 'I can still hear the awful clamour as the French came to under our lee quarter.' Then, with eyes tight shut, the better to play the scene out again in his mind, he continued.

'In the midst of this desperate onslaught, I saw Captain Bodie mortally wounded. Even today I can still recollect the anguish on his face. Sinking to his knees on the deck, pierced by an enemy's sword, this man who had so cruelly abducted me was now in mortal anguish. For all the suffering which he had caused me, I cannot say that the sight of his brave death was anything other than painful to behold. But even this sword thrust was not enough to quell the indomitable spirit that animated him. Our captors' cruel barbarity exceeded all our fears when, as this already mortally wounded man attempted to stagger upright, he was again attacked.'

Here William took a pause from his painful recollections.

'A ferocious thrust from a lance met Captain Bodie with such force and skill that its shaft went clear through his body and into the deck itself. His cries, as in his wild efforts to release himself he turned on the lance, were pitiful to hear. A too cruel and ironic fate had consigned him to an ignominious death; made him the quarry until, like the very whales which he had hunted, he was struck down and pierced through his heart.

'But a still more heart-rending shock was to come. As his life's blood gouted onto the deck the Captain raised a feeble hand towards his brother, perhaps in supplication, perhaps in warning. William Bodie, who obviously perceived that defence was ineffectual, was then on the quarterdeck in the very act of lowering our colours. His brother's anguished gesture turned our captors' attention to this new adversary and they now unleashed their bloodied frenzy upon him. I watched in helpless horror as they shot this one true friend dead where he stood. For a single moment it seemed that his eyes searched out mine. He had seen his brother killed before his eyes and his last thought was surely for my own safety. A blackness rolled over me. I heard the sound of a body hitting the deck but felt nothing as I lay there swooning, and I remember nothing more until I came to in the hold below.

'The privateers commanded us to go below where we were barred down and kept secured. Having done this, they made sail on the *Frederick* and steered us in convoy before the wind towards the coast of France. During this confinement we became seriously concerned for our future destiny. Some deeply lamented the long anticipated separation from home and friends which would unavoidably take place in consequence of an expected confinement in a French prison. Others thought more of the privations and sufferings which their destitute families would have to endure during their imprisonment. We were prisoners of war and the *Frederick* a captured prize.

'Yet kind fortune had unrevealed and unmerited favours reserved for us. The *Frederick* and her captors continued their course for over a week and early one morning arrived off the mouth of the Bay of Bologne. The Frenchmen had obviously intended to sail into the harbour before the break of day, if possible, to escape any English cruisers. But in consequence of the darkness of the night, they had been compelled to shorten sail and, the fury of the wind having abated towards morning, they were frustrated in their designs. Daybreak overtook them while they were still on the ocean and thus they were exposed to English retaliation.

'The day was breaking when the ships commenced entering the harbour. Not having yet discovered any of the English cruisers the Frenchmen were in great hilarity, thinking themselves in perfect safety with their prize. But the scene soon changed; those who had been rejoicing and quaffing deeply of the cup of victory were soon brought to drink just as deeply of the bitter cup of adversity. Most of this was only revealed to me later, confined as I was below and therefore ignorant of much that was happening. The first intimation of trouble on deck came when all noise suddenly ceased, only to resume with unintelligible shouts in French. "*Mon Dieu! Les Anglais!*" resounded all over the ship. The privateers' consternation on sighting an English brig of war could be imagined and thereafter a long and sustained chase began. This only concluded when we heard the report of a mighty cannon, which led us to believe that our redemption was nigh.

'Soon after this we heard the sound of a speaking-trumpet delivering a warning in English and immediately afterwards the shrill sound of boatswain's pipes. These were followed by the roll of a drum and a huge concussion on the lee-quarter which shook the vessel fore and aft. This was succeeded by a great noise on deck as men ran everywhere in confusion and panic. English voices, commanding, were now distinctly heard ordering the dismayed Frenchmen to the quarterdeck. This heaven-sent reversal to the *Frederick*'s misfortune was enough to set some of our stoutest men to breaking out of the hold. As they stumbled on deck it was to witness a scene of confusion, with two mighty British frigates towering alongside.

'The privateers made some small resistance when boarded, rather than submit to their fate, and in the ensuing mêlée I was seriously injured. The sight of the huge frigates quite overwhelmed us and the gallant efforts of our crew to secure a rescue encouraged me to advance too closely into the fray. For my youthful bravado I was to pay dearly, with a sword thrust through the side and a vicious injury to the hip. Although my pain and anxiety confused my senses, I know it was but a short time before I was being cared for by a doctor from one of the frigates. I will never forget the first look on his kind face; it was one which was to become so familiar to me during the next four years.

'Having thus obtained possession of the *Frederick*, the English commanding officer immediately ordered the French crew to be confined. Thus, in a few minutes, our situation was greatly changed. Our former captors were now prisoners, while all below were liberated. Hearts which had previously been heavy with adversity now swelled with gratitude.

'The doctor himself, Daniel Querrier, in a later petition to the Admiralty described our meeting, saying that my "extreme youth and destitute and forlorn condition excited compassion and humanity."'

STRANDS OF RUMOUR

These then were the first and second acts of my unfolding saga which John Campbell pieced together for Governor Macquarie. The story, even without the tutor's shaping hand, possessed an unstoppable drama whose colourful incident engaged the Governor's most sympathetic attention and unfeigned interest.

Some short time after compiling his text, John Campbell's suspicions about the responsibility and involvement of Captain Bodie in my abduction were unexpectedly and shockingly confirmed. It happened that his path crossed that of the Rev Samuel Marsden, a much respected colonist and sometime visitor to Government House. Knowing the Reverend's interest in philanthropic works, Campbell acquainted him with his recent posting as scribe and tutor. Only after he had brought the story full circle and explained his perplexity about my first step into the unknown in 1811, did he actually name the *Frederick* and its Captain.

It was only then that his avid listener interrupted the narrative, something he had previously refrained from doing. 'But surely this is the same Captain Bodie I came across five years hence? It was a distressing case, serious enough to warrant immediate action rather than await the return of the Governor from a country visit.' It was obvious that the memory still stirred his sympathy. 'I remember it clearly. I had only been back a year from England and newly installed as Principal Chaplain to the Colony.'

Here, at last, my tutor knew that he would finally learn more about my cruel removal from home and family. In response to his urgent entreaty the reverend gentleman continued: 'It was sometime late in 1810, the year before your story began. I was one of three people, including the Captain's brother William, who were appointed to take charge of Bodie and his property.'

Now it was Campbell's turn to interrupt. He asked in some bewilderment, 'Take charge? Why ever would you be asked to take charge of the man? Was the Captain a criminal?'

Sadly, the Reverend Marsden shook his head. 'No, no. Never a criminal but a most unhappy man.' And here he sighed deeply and, after a short silence when he appeared to be wrestling with his recollections, said, 'There will be notes in the Governor's private papers to which, I presume, you could be given access. But I really do not feel that I can discuss such a personal and private matter any further.' And with this abrupt and utterly bewildering finale the Reverend turned and quickly left. It seemed almost that he fled from any further entreaties to continue his story.

This left Campbell even more mystified. There was, as he had suspected, much about the life of Alexander Bodie that had yet to be disclosed. Strands of rumour had attached themselves to this man like seaweed to a boat, and the erratic voyage that his life became trailed disaster in its wake. A new zeal to uncover the past of this man soon led to Government House. What Campbell found there was confirmation of what he already suspected.

The fifth Governor of the Colony of New South Wales, the Hon Lachlan Macquarie, arrived only the year before William's abduction. Thereafter Sydney-town was transformed into something which Captain Phillip would hardly recognise or credit. It was no longer a military camp and free settlers justifiably took pride in an increasingly ordered streetscape and an evolving town plan.

There was an unprecedented flood of general orders and proclamations from the Governor's office which were printed by an overloaded *Sydney Gazette*. It was also significant that the town common was reborn as the more grandiloquent Hyde Park. In short, the new Governor combined a welcome paternalism with popular town planning. To his delight, the turnpike road to Parramatta had been finished in November 1810 and he and his wife undertook a vice-regal visit to the outlying settlements. Their splendid entourage included a party of dragoons, Bengal tents coloured their equipage and their progress amazed and even cowed many along the way.

But it also meant that authority had been left in the hands of the Lieutenant-Governor, Maurice O'Connell. And it was while he briefly substituted for the peripatetic Governor that a most difficult duty fell upon him.

A whaling ship out of London which carried general merchandise and sailed under the command of Alexander Bodie had arrived in Sydney Cove on 21st October 1810. Almost immediately the Lieutenant-Governor was approached by a local merchant, Robert Jenkins, whose business dealings obviously brought him into close personal contact with the Captain. This contact had alarmed him enough to request that the Lieutenant-Governor 'appoint a Medical Committee in order to Examine and enquire into the State of Health of Mr Alexander Bodie'.

A medical board was quickly convened. The Acting Principal Surgeon of the Colony [and soon-to-be Police Magistrate] D'Arcy Wentworth, together with the Surgeon and Assistant Surgeon of the 73rd Regiment, met on 19th November. Their finding was unequivocal: 'In conformity to the general Order of yesterday, we the undersigned met in professional consultation to examine into the State of Health of Mr Alexander Bodie, master of the Ship *Frederick* of London, and find him labouring under Serious Mental Derangement, which renders him totally incapable of performing the Duties of his Situation.'

Once this lamentable verdict had been delivered, events moved quickly. The very next day Bodie's command had been taken over by the Colony's most respected whaler, Eber Bunker. Within a week, three respected colonists were appointed 'to take Charge of the Person and Property of the said Alexander Bodie' with the injunction 'to perform Such acts therein as may be most conducive to the Restoration of the right Exercise of his Mental Faculties'. These three were Bodie's colleague Robert Jenkins, the Rev Samuel Marsden and the Captain's brother, William Bodie.

After finding such detailed information confirming Captain Bodie's derangement, John Campbell was frustrated in any further attempts to discover how this malaise manifested itself. Was the Captain prone to violent fits, to erratic acts which endangered his crew and the ship, and had he succumbed to delusions that terrified all at sea? Was the madness sustained or episodic and what prognosis did the medical committee put on the Captain's possible recovery?

To all these questions he drew a complete blank. The affair had been settled with great speed and a minimum of publicity and it appeared that not a breath of scandal had penetrated into the wider but close-knit community.

Before the month was over, Bunker had sailed the *Frederick* out of Sydney for Hobart-town and the Fishery. The ship was gone for almost nine months and arrived back in Sydney on 21st August 1811. Less than two months later the whaler began its long voyage back to England with Captain Bodie reinstated at the helm. Did the medical committee believe that in just under one year Captain Bodie had been restored to his former competence? Surely events thereafter proved that any such diagnosis or optimism had been completely mistaken?

My abduction thus became yet another instance of Bodie's continuing derangement. Any optimistic diagnosis about the Captain's restoration was entirely unfounded and at least one person accepted this sad conclusion — William Bodie, the Captain's brother. In confirmation of this assertion, John Campbell was convinced that when Bodie's brother joined the crew for the homeward voyage of the *Frederick,* he sailed with the obvious role of keeping watch over someone he knew was still labouring under mental difficulties. And sadly, this caring gesture by one brother to another was to culminate in the death of both off the west coast of Africa a year or more later. For surely it was the ultimate manifestation of the Captain's 'Serious Mental Derangement' that steered the *Frederick* away from the protection of a convoy and into the hands of murderous pirates?

Faced with the sad facts, John Campbell could not but wonder at the symmetry he now perceived in my ordeal. The colonial authority asked by the Admiralty to provide protection and assistance to me had been the same authority which, completely innocently, had set a fatal conjunction in motion five years earlier.

THE WHALESHIP *ESSEX*

Once Mr Campbell's Account of my exile had been compiled I finally felt that I could begin a renewed existence. It was remarkable how quickly the passing weeks wove me back into everyday life.

I had now been home and reunited with my family for just over four years. One day — it was the middle of 1821 and a day I had good cause to remember — I felt Father's kind hand on my head. There I was, surveying the harbour from the eyrie on our rooftop. It had always been my favourite vantage point and when I was not out taking long rambles across the town or around the harbour foreshore I was content to retreat up there.

By now the family had become accustomed to my solitary nature and my need for quiet reverie. Without a word, Father sat down next to me and together we looked out over the majestic expanse of water. There was no need for speech. Finally, however, Father broke the silence with a question.

'William ... what do you contemplate, sitting up here all alone? Your mother and I know that you are content in your own company but we can see you are yearning for something. Do you know what it is? Can you tell us?'

These were difficult questions to answer, for if I divulged what was in my heart I knew that my parents would be deeply hurt. For the truth was that after unprecedented exploits and a life of extraordinary adventures at so tender an age, I was not content to become the settled son my parents so clearly desired. They had been delighted that I had taken so eagerly to the tuition provided by the Governor's secretary, Mr Campbell. In truth, though, it had hardly been a chore. Dr Quarrier's patient lessons had already given me a taste for the world of books and learning and my final promise to him had been that I would not neglect my schooling.

And thus I found myself striving to satisfy my actual parents as well as one who had stood in that caring capacity and still claimed my thoughts and affections, even though he was back in England. So it was that for all their sakes I had even contemplated becoming that most sedentary of creatures, a writer. Mr Campbell had introduced me to an acquaintance of his, the editor of the *Sydney Gazette*, Robert Howe, and for a time I even entertained the notion of a literary life. I had fallen into the habit of keeping a regular journal and its immature offerings won some encouragement from the *Gazette*'s editor.

Perhaps writing and the world of the imagination would satisfy the taste I had acquired for novelty and change of scenery. Mr Campbell had long since completed his duty in committing much of my strange story to paper, but in the process he had become such a genuine friend that he was still an occasional visitor to our home. One morning at the very beginning of June 1821, a day not easily forgotten, he arrived even as breakfast was still progressing.

'Mr Campbell, oh my goodness ... well ... you ...' my mother flustered around this unexpected and early arrival. She had never quite become accustomed to the honour of our visits from the Governor's secretary. Although he did his best to put her at ease, she was always nervously looking for something to do as she danced attendance on his august presence. For this reason, and because I was sure he had some purpose in mind, I quickly finished up and suggested that we might enjoy the mild winter's morning and take a walk together.

His purpose was quickly uncovered. It concerned our mutual friend at the *Sydney Gazette*. This literary idol of mine — no matter that his efforts were published in our colonial backwater far from the world's stage — had invited us to accompany him to what he believed would be a rare interview and an extraordinary coup for his newspaper.

The much respected Thomas Raine, Captain of the *Surry*, had arrived in Sydney Cove with three rescued survivors from an American whaleship, the *Essex*. Apparently, just as Captain Raine was on the eve of leaving Valparaiso for Sydney he fell in with a vessel which had rescued two shipwrecked whalers. These men had told of three others left behind on a remote island and, as Captain Raine was bound somewhat in their direction, this good Samaritan resolved to attempt their rescue.

That his mission had been successful was cause enough for the *Gazette*'s interest. But what piqued our curiosity even more were the wild stories already circulating around Sydney-town from those of the *Surry*'s crew who had been ashore. Reports of a vengeful whale literally tearing the *Essex* to pieces seemed scarcely credible. Worse still, it was said that of ten who had originally escaped in one small whaleboat, two alone had survived their ordeal and that only by resorting to that most repulsive of extremes — cannibalism!

I heard less dramatic but more reliable information about the arrival of the *Surry* from my companion as we neared the harbour. Captain Raine had been at Chile loading a cargo of wheat urgently needed for our hungry colony and had left there on 10th March 1821, determined to effect the rescue of three reported castaways.

Together with our mutual friend we were welcomed aboard the *Surry* by its Captain. Before he introduced his three rescued whalers, who had languished for over three months on their island, he assured us that they had recovered their strength and accustomed cheerfulness and were honoured to submit to an interview. Captain Raine also felt it incumbent upon himself to immediately disclose the awful truth behind the wild conjecture he knew was circulating. He had no wish to pre-empt the survivors' revealing their own saga but did narrate the sorry story which had been revealed to him in Valparaiso.

The *Essex*, of 260 tons, sailed from Nantucket in mid-August 1819 under the command of Captain George Pollard, Jnr. Its whaling voyage to the South Seas had procured 750 barrels of oil before disaster struck. On 20th November 1820 they were among many whales. The prospects for filling even more barrels seemed excellent when one of these whales, of the largest class, struck the *Essex* and knocked off part of the false keel. The animal remained for some time alongside, endeavouring to clasp the ship with her jaws, then went away about a quarter of a mile and, turning suddenly, came at the ship with tremendous velocity, head on.

The *Essex* was going at the rate of five knots, but such was the force when the whale struck the ship that it turned so violently that the sea rushed into the cabin windows. Every man on board was knocked down and, worse still, all the bows were stove in completely. A very few minutes later the vessel filled and went on her beam ends. At this unhappy juncture the Captain and second mate were still off in whaleboats, with both actually fast to a whale. However, on beholding the catastrophe they immediately cut loose from the fish and made for the ship. The whale, meanwhile, remained in sight and its unprecedented action aroused fears that it might return. But perhaps it had received an injury in its violent collision, because the *Essex* was mercifully spared another visit.

By cutting away the masts the vessel was righted. The upper deck was then scuttled, some water and bread were procured and twenty men abandoned ship in three open whaleboats. In expectation of falling in with some vessel they remained over two days by the wreck, making sails, but at length were compelled to abandon it.

On 20th December they made landfall on an island, where they remained for a week, but with the island affording little or no nourishment the majority resolved on venturing for the coast of South America — leaving behind the three men now on board the *Surry*. The continent was seventeen hundred miles distant and in their destitute condition they could scarcely expect to reach land. Their hope, rather, was still directed at the possibility of falling in with some vessel.

One of the three open whaleboats carried the Captain, a cabin-boy and four men. Yet when they fell in with a whaler from America, only two survivors remained. The horrible account given by these two was appalling in the extreme. They had been ninety days at sea and had experienced the most dreadful of all human vicissitudes; from the extremity of hunger they had been reduced to the necessity of killing and devouring certain of their number in order to sustain a wretched life that was hourly expected to be terminated. Lots had been drawn and the losers had been sacrificed to afford sustenance to those that remained. And on the very day that a ship encountered them, the Captain and the boy, who alone remained, had just drawn lots — determining that the poor boy should die. But, providentially, a ship hove in sight and took them in and they were restored to human companionship.

Doleful in the extreme as it is to hear such things, and painful as it is to relate them, it was nevertheless asserted as fact by Captain Raine that the fingers and other fragments of their deceased companions were in the pockets of the Captain and boy when taken on board the whaler.

'It was at this stage,' said Captain Raine, 'that I became opportunely acquainted with these awful circumstances and of three men still languishing on their lonely island. As it was no great distance out of my track from Valparaiso to New Holland, I determined to be instrumental in restoring them to society. By mid-April we had arrived at an island, but since we were not certain if this was our landfall we discharged a gun to arouse the attention of anyone on shore. Within a short time three unfortunate men were seen to issue from the woods. I gave orders to lower the boats and took one myself.

'On approaching the shore it was found not only dangerous but utterly impracticable to land, of which circumstances we were informed in weak and tremulous voices by the almost starved creatures themselves. One poor fellow summoned up courage to plunge into the waves and reached my boat. After we warily backed in as near the rocks as I thought possible, amidst a heavy surf, all three proceeded in getting on board, much bruised and lacerated.'

Here Captain Raine paused, shaking his head in sad recollection of a moment fraught with such emotion. 'Suffice it to say,' he continued, 'that every precaution and tenderness was observed and, in a very few days, all three soon recovered. They were even, shortly, enabled to assist my own crew in their daily tasks. They declare their names to be Thomas Chappel, William Wright and Seth Weeks.' At this introduction three young men came forward.

There was nothing in their outward appearance to distinguish them from the rest of the crew and we listened eagerly to their account as Chappel, who'd been second mate on the *Essex*, spoke up. 'We three chose not to trust ourselves again to the open sea and yet the leave-taking of our braver companions, after one week on the island, was heart-wrenching. We separated with mutual prayers and good wishes and stood vigil long after their boats had passed from sight.'

The haunting recollection of this parting was deeply affecting to all of us. Mr Howe, indeed, listened spellbound [and relied later on his prodigious memory to recreate the story]. It was Mr Campbell who asked in amazement, 'What were the prospects of life on this remote island? Captain Raine has told us that it was almost entirely destitute of the necessaries of life and that no fresh water could be discovered.'

The second mate continued his doleful tale. 'Our prospects were indeed gloomy. We continued to search for water, in vain. Then we tried to dig a well, but without success, and all hope seemed at an end when, providentially, we were relieved by a shower of rain. For food we relied on small birds which we caught at night. We climbed trees in which they roosted and by these severe exertions obtained a scanty diet, hardly enough to support life. There were also trees that bore a small berry which gave us some little relief.

'We also searched in vain for shell-fish. Although we saw fish and a number of sharks, without fishing tackle we had no way of adding this bounty to our meagre supply. Once we found several turtles and even succeeded in taking five, but before we had quite finished one, the others had become unfit for food.'

I was beginning to wish that I had not hurried my breakfast so much when Mr Campbell asked, 'Did you never find a spring on the island?' Thomas Chappel shook his head. 'We suffered most from want of water; our only supply being from what remained in holes among the rocks after passing showers — and sometimes there were five or six days without any. On these occasions we were compelled to suck the blood of the birds we caught. At first this allayed our thirst in some degree, but we soon found ourselves much disordered thereby.

'We continued in this sorry state for four months in anxious expectation that some vessel might touch at the island and we prayed earnestly for our deliverance. We had become increasingly feeble and after so long a time feared that our more intrepid shipmates had perished at sea.

'Then one morning our attention was aroused by a sound which we thought was distant thunder until, on looking towards the ocean, we saw a ship in the offing which had fired a gun. Our relief was unbounded; we signalled from the shore and then fell to our knees thanking God for our deliverance. Though life may be across a stormy and tempestuous ocean, yet a true believer proceeds with confidence, assured that he shall reach his safe harbour in safety.'

Here our narrator finished his story, but so moved were we by his ordeal that it was some time before anyone spoke. The extreme dangers which these three had undergone and their providential preservation had made a deep impression on us. Mr Howe was suddenly roused to his note-taking again and had questions about various particulars for his forthcoming article, which would be sure to move even the basest felon in this convict colony to a pious exhortation. This scarcely credible story, so full of violent incident and cruel outrage, was given the almost unprecedented distinction of filling two complete columns in the *Sydney Gazette* on 9th June 1821.

It was finally left to Mr Campbell to echo our opinions when we took our leave from the *Surry*. 'Mr Chappel,' he said, shaking the man's hand, 'nautical history presents us with many instances of unexpected dangers and remarkable preservations, but few are more worthy of notice and compassion than what you have related. I intend to apprise His Excellency the Governor of your situation and the extent of your sufferings, and I trust that you and your companions will soon enjoy his philanthropy.'

A GENERAL EXODUS

My recollection of this unique and moving meeting with the survivors of the *Essex* remains undimmed after thirty-five years. In all the years I have since spent whaling I have encountered no event more gruesome or notorious than the fate of this ship. I give thanks that no personal experience of mine at sea ever reduced me to the desperate extremes faced by this crew in pursuit of their trade. But their harrowing story left a shadow over our fraternity which has never been forgotten.

With the bravado of youth and a belief in my own personal invincibility, I admit, I once saw the fate of the *Essex* as a great adventure and an exciting test of mettle under extreme conditions. And hadn't I already survived equal dangers against all odds? Nothing would have deterred me from pursuing a life at sea. Now, however, my youthful reaction is overlaid with a lifetime's experiences. A generation later I see it differently, with more wisdom as well as great sadness. Now I see the suffering on the *Essex* as an indictment of the Devil that lurks in all men, ready to be unleashed with savage violence.

But these mature opinions, now dreadfully coloured by my son's brutal murder and my own dark reading of mankind, were then many years of sailing into the future. And even if I had entertained any such gloomy thoughts in 1821, the rush and excitement of the preparations to farewell the Governor would soon have set them to one side.

The day I recall most vividly during these preparations, and one I was able to enjoy from a position of privilege, was 1st December 1821. On that day, in pursuance of Government and General Orders, an exodus from Sydney-town westward to Parramatta Market-place occurred. This memorable event was the Annual Meeting of the Native Chiefs and their Tribes with our Governor, and Mr Campbell had invited me to accompany him in the official party.

At the time he also explained the origin of this extraordinary confluence. 'These meetings started in December 1814, a few years after you were abducted, and have since proved to be an anticipated part of our colonial calendar.' Both my father and myself were astonished to hear that over three hundred Natives from different Tribes were expected to attend. At this my father expressed some alarm. 'But surely such an intermingling of Tribes is fraught with differences likely to be exhibited in some act of warfare,' he remarked. 'I recall the sad fate of our founding Governor whose own attempt at a friendly meeting had such unforeseen results.'

Mr Campbell hastened to reassure us. 'All that was a generation ago. At today's general exodus there will certainly be some Natives who have never had contact with Europeans before. Even so, Governor Macquarie's distribution of blankets and food has never been mistaken for anything other than a gesture of goodwill.'

I was further told that these annual meetings had become popular spectacles with the general public who were much attracted by the novelty of the scene. 'You will find, William, that the Natives who attend this meeting from outlying districts are vastly different from those forlorn wretches you see around this town. They will assemble manfully with their respective families and display a pride and a bearing that marks them out as the original inhabitants of this ancient land.'

Thankfully my knowledge of the Natives had not been limited to those sorry examples that habitually roamed through the Town and had been degraded by our 'superior' civilisation's vaunted advantages of rum and tobacco. Perhaps these others which I had seen were more at ease with children and young people in the settlement, and thus I had formed my own opinion of them and their unrestrained and usually warranted amusement at our expense.

Today I believe that if an accurate observation and a quick perception of the ridiculous be admitted as proof of natural talents, then the Natives of New South Wales are by no means deficient. In proof of this I recall that there had indeed been times when their mimicking of the oddities, dress, walk and looks of ourselves was so exact as to be a kind of register of our oddities rather than theirs.

If there was anything peculiar in any of our countrymen, any officers in the Corps or even among the convicts, any cast in the eye or hobble in the gait, any trip or strut, stammering or thick speaking, then the Natives caught it in the moment and represented it in a manner which rendered it impossible not to recognise the original. They were, moreover, great proficients in the language and Newgate slang of the convicts and, in case of any quarrel, were by no means unequal to them in the exchange of abuse. I can see that this attribute would earn them the rancour of their targets.

The Governor scheduled these annual meetings in late December or early January, which is the height of summer in the Colony. Yet, even so, an excited concourse of Sydney's civilised elite, together with the general populace, were all intent on braving that heat. All were eager to observe the exotic contrast which the day's meeting would provide. I listened to Mr Campbell's words as we made the slow journey on horseback along the turnpike road to Parramatta.

'Those Natives who resort to the cultivated districts of this settlement, although prone to a seeming indifference as to their future means of subsistence, yet in general are of free and open dispositions. Indeed, it is my opinion that they are devoid of that designing trick and treachery which characterises the Natives of New Zealand and those of the generality of the islands in the South Seas.'

This considered assessment led me to recall my opinion of the Natives I had seen in Timor some six years previous. Although this was my only real contact with a totally foreign population, I was not naïve enough to believe that my experience there in any way prepared me for today's meeting. I listened further as my travelling companion's words gave me his opinion of the Governor's Native guests.

'The principal part of their lives seems wasted in wandering through their Native woods, in small tribes of between twenty or fifty, in quest of the immediate means of subsistence. They make opossums, kangaroos, grub worms and such animals and fish as the country and its coasts afford the objects of their fare.'

Out of respect for Mr Campbell I did not comment on the criticism explicit in his description of the Native's mode of existence as 'wasted'. It seemed to me, perhaps in my young ignorance, that such a simple existence, at ease in their land, was something to be envied. But then I could hardly question the opinion of the Governor's secretary.

The morning heat pressed down on our dusty journey and even the shrill intrusion of the cicadas' incessant noise could not prevent me from falling into a mid-morning stupor. Shortly before eleven in the forenoon I was roused by our approach to the Market-place and the excited sound of a large crowd. The Governor was accompanied by the Lieutenant-Governor, most of the officers — civil, military and naval — together with a vast assemblage of the most respectable inhabitants of the colony. A grand circle was assembled with chairs provided for the Chiefs of each Tribe represented. I was told that some of these Tribes had travelled from beyond the Blue Mountains to be present and my eager gaze, now thoroughly alert, quickly distinguished these intrepid travellers from their more local brethren.

Their general appearance and peculiar style of ornament at once singled them out from those inhabiting the country on this side of the Rivers Hawkesbury and Nepean. They had decorated their hair with white feathers and had the teeth of wild animals suspended in rude festoons over their foreheads. Their bodies and faces were painted with red and white ochre which rendered them singularly wild and outré. Although this was undoubtedly the first time that some of them had held any intercourse with white men, they still displayed, to an enviable degree, a proud consciousness of their station.

In contrast, the local Natives had left their adornment until this morning and their preparations were an exciting prelude to the day's unfolding. Their young women were employed with all their art in painting the young men, who were chiefly adorned with streaks of white, done with pipe-clay and in different forms according to the taste of the man himself or that of the partner who adorned him. Truly no fop preparing for an assembly was ever more desirous of making his person irresistibly appealing.

This paint could not be applied without a little moisture and we were all astonished to see the method with which this toilette was accomplished. The lady, as she drew those marks on the face and chest, would spit on the body to make the clay mark the stronger. Perhaps it was a sign of special esteem to be so treated and certainly not one of the warriors was in the least perturbed by this custom.

There were other Tribes from the north and south who had travelled a distance of upwards of one hundred miles. Mr Campbell was busy taking notes and making drawings. 'It is a gratifying prospect,' he said, 'to behold so many various Tribes assembled together on a social and friendly footing and disposed to forgo those customs which interdict them from travelling over the allotted ranges of others.'

I must admit that I was paying scant attention to his comments. I was more concerned with enjoying this extraordinary spectacle, following the shy gaze of the Native children as they took in the regal splendour of the Governor's dress and, I admit, in spite of Mr Campbell's stated confidence, somewhat apprehensively appraising the Natives' fearsome appearance.

Now, all being ready, the dance, or corroboree, began in earnest. The first to start were a few young boys who were then joined by men and women, chiefly the former, until their number amounted to several dozen. Their dance was truly wild and savage, yet, in many parts, there appeared order and regularity: one man would frequently single himself out of the dance and, running round the whole of the performers, sing out using loud expressions which we could not understand. Perhaps these were intended to urge his companions on to greater effort and display. Then he would rejoin the dance. Thereafter certain parties would alternately lead forward to the front and there exhibit, with their utmost skill and agility, all the various motions which, to their opinion at least, seemed to constitute the principal beauties of dancing.

One of the most striking of these displays was that of placing their feet very wide apart and, by an extraordinary exertion of the muscles of the thighs and legs, moving the knees in a trembling motion such as none of us could ever mimic. Our sorry show later in poor imitation of these muscular exploits seemed to prove that it required much practice to arrive at any degree of perfection in this singular motion.

There appeared a good deal of variety in their different dances; in one, for example, they paired themselves and frequently danced back to back. They then changed suddenly and faced each other, or sometimes all the performers sat down on the ground with their feet under them and, at a particular word or order, raised themselves up without any assistance from the hands. Now they ran back in direct rows, then advanced in the same order; again they would form a circle with some distinguished person in the centre, and sometimes the whole of the performers would appear waving a green bough in their hands in a conspicuous but, to us, unintelligible display. Whatever their purpose, however, it was obvious that an animating spirit of fraternity and social amiability permeated the whole.

In all the different figures which they performed, I observed that they generally finished with certain numbers of their principal dancers advancing to the front and repeating their favourite part of the dance, namely the quivering motion of their knees. I concluded that these were such as had great confidence in their own skill and no doubt were vain enough to outshine in their ability the rest of the company.

The music which accompanied these exertions involved the use of two sticks of very hard wood, one of which the musician held upon his breast in the manner of a violin and struck with the other in good and regular time. The performer, who was a stout and strong-voiced man, sang the whole time with the assistance of several young boys and girls. These few sat at his feet and by their manner of crossing their thighs made a hollow between them and their belly upon which they beat time with the flat of their hands. He, poor man, stood up the whole of the time and seemed to me to have the most laborious part of the performance.

The dancers very frequently, at the conclusion of parts of the corroboree, would apply to us for our opinions, or rather for marks of our approbation of their performance. These we never failed to give, often repeating the word *boojery,* which signifies good, or *boojery carib-berie* — a good dance. These signs of pleasure in us seemed to give them great satisfaction and generally produced more than ordinary exertions from the whole company of performers in the next section of their dance.

Then, before a distribution of blankets and food began, the Governor made a short speech. 'It pleases me to say,' he began, 'that our Native guests here today, who dwell near Sydney or the other principal settlements, live in a state of perfect peace, friendliness and sociality with our settlers.' At this, a gruff exclamation came from a rough-looking fellow nearby. It was enough to convince me that the vision from Government House, although well meaning, was most probably uninformed and overly paternalistic. It failed to appreciate the Natives' radical differences from ourselves and overestimated the settlers' belief in a peaceful co-existence.

'It seems,' the Governor continued, 'only to require the fostering hand of time, gentle means and conciliatory manners to bring these unenlightened people into an important degree of civilisation.' Once again my less charitably disposed neighbour snorted his rude opinion of the Governor's optimism. And this time his opinion was validated by approving nods from a group of settlers whose sour and disapproving looks placed them so obviously at odds with the sentiments of those in our party.

Unaware of such dissenting opinions, the Governor continued: 'My preliminary measure in offering them the benefits of civilisation has been the establishment of the Native Institution here. It has been designed to bring up the youth of both sexes in habits of industry and decency.' And at this juncture his gaze fell upon a group of a dozen or more Native children whom I had noticed earlier. This separate assembly was most carefully dressed, not in their own clothes of skins but in English dresses and trousers which they had been shyly showing off to admiring relatives.

'I am informed by their Tutor that he is highly gratified by their progress in reading and writing, and particularly with the attention the little females have paid to their needlework — thus bringing to the expanded mind of the Philanthropist and the Christian the peculiar advantages of this Institution.'

After His Excellency had conferred some badges of honour and merit on the Native Chiefs, these children were introduced into the circle, where they were shown to their families together with specimens of their progress in the schoolroom. Their drawings seemed to delight the elder Natives beyond anything yet presented. They manifested their approval by bursts of loud laughter and by leaping upward with wild gesticulations.

The Governor advanced once more to deliver some final words, whose pertinence and pathos seemed particularly to affect his loyal secretary. 'As this is the last such assembly I shall attend before I sail for England, I trust you will allow me some more personal comments.' These words, which reminded us of his imminent leave-taking, brought a closing sadness to the festivities and a realisation that an era was truly closing.

'My most fervent prayers will always be offered for the welfare and prosperity of this Country and for the happiness of its Inhabitants; fondly and confidently anticipating that, in less than half a century hence, it will be one of the most valuable appendages belonging to the British Empire. I shall not fail on my return to England to recommend in the strongest manner I am able, to my Sovereign and to His Majesty's Government, their early attention to the amelioration of this valuable rising Colony and to extend to it their paternal support and fostering protection.

'The length of time I have governed this Colony, the progress it has made in improvement during my Administration, and more especially the fond recollection of my only surviving Child being born in it, all combine in attaching me most strongly to its interests. I shall not fail to cherish the same sentiments of attachment in my son who, although yet so young, feels and already expresses the strongest affection for his Native Australian Land. With these emotions and wishes deeply engraved in my heart, I now bid you all — Farewell!'

When these ceremonies had passed over, the whole of the Native party was regaled with roast and boiled beef, forty loaves of bread, tubs of potatoes, puddings and massy plumcakes, as well as a hogshead of watered grog. It was soon early afternoon and everyone was sated with their feasting and drowsy with the heat. The children from the Native Institution were the first to retire. They left for their school in the same order in which they had arrived, marching in pairs, the females foremost, all under the direction of their Tutor.

We had been granted a vision of social intercourse and amiability that day which truly inspired hopes for a peaceful co-existence in the future.

THE GOVERNOR'S FAREWELL

When the sun rose on Tuesday 12th February it brought a bright summer's day to Sydney — one whose cheerful mien was distinctly at odds with that of most of the population. This was the day Mr Campbell had long been dreading, a day when this devoted secretary, who had served the Governor for his entire term of office, must now say farewell. And Mr Campbell's sorrow was felt generally throughout the Town. An immense concourse manifested, by their melancholy looks, a severe and undisguised regret at this parting.

It is true that there had been numerous events by which the Colony attempted to show its gratitude to this great man. Only the week before on 31st January, which marked the sixty-first anniversary of his birth, he had been treated to three distinct celebrations. He had breakfasted at Elizabeth Bay with a group of Aborigines, lunched with prominent emancipists and later dined with a select group of free settlers. In this way the fundamental divisions in Sydney-town society — Native, ex-convict and freemen — were all accorded due regard, even if the Governor had been unable to reconcile their differences during his term of office.

At daybreak on this final day the Governor dressed and handed over his last official papers and keys. Then at noon, accompanied by his wife and young son, he left Government House just as he had arrived over twelve years earlier, in full panoply. Here he was attended by the principal civil and military officers and by many of the gentry. Together with this solemn entourage he walked across the Domain, through lanes formed by proud schoolboys and by soldiers of the 48th Regiment, to a barge moored off Bennelong Point. This leave-taking was thus an occasion for much public ceremony and revived memories of a nabob's procession which I had witnessed in Timor.

Amid all this spectacle it was amusing, and yet oddly touching as well, to see that the Macquaries were even taking their favourite old cow home with them. No doubt he would be as coddled as 'Sultan', the Governor's horse, who also had a veritable farmyard of companions for company. But most colourful of all was the raucous assemblage of native birds and animals also making this voyage: seven pet kangaroos, five emus and seven black swans along with native companions, white cockatoos, parrots and pigeons.

The band played 'God Save the King', the Regiment presented arms and the officers all saluted in honour of their departing Major General. Half an hour later His Excellency and family embarked on the barge which took them out to the *Surry*. Since early morning most of the vessels in our harbour had been crowded with onlookers and had been decorated honorary to the occasion. As soon as this barge with its illustrious burthen retired from the shore, a salute of nineteen guns was fired from Dawes's Battery.

Launches, barges, cutters, pinnaces and wherries all crowded the harbour, their occupants determined on catching a last glimpse of their departing Governor who, for some time, stood proudly tall and kept bowing adieu as he passed. Adverse south-westerlies, however, obliged the *Surry* to anchor in Watson's Bay, off South Head. It was almost as if Nature herself had conspired to prevent the Macquaries from leaving the Colony, for it was not until the morning of Friday the 15th that the ship was able to weigh anchor.

By then a fine breeze had at last sprung up, west by sou'west, to carry the *Surry* on her voyage south of New Zealand and via Cape Horn. By late that afternoon the Macquaries' parting view of the Colony would have been the tip of the lighthouse which had so recently been finished at the entrance to the harbour. And then ... the open sea.

And I must admit that it was this welcome thought of the open sea which stayed in my mind more than any other emotion. Unlike many that evening, who still wore the sad signs of parting, I could not help but dwell on the uplifting prospect of the long voyage which was now ahead of the Macquaries. My own yearning to be at sea again far outweighed any sadness. More than anything I missed the comforting dialogue between the creaking timbers of a ship and the answering voices from wind and water.

In saying goodbye to an era, I finally realised that my own future and my heart's desire lay out there on the limitless ocean. If I had not been frightened off by an insane sea captain, the attack of pirates or stories of vengeful whales and cannibalism, there was little chance that anything could deter me from a determination to return to sea. I was now eighteen; it was time to make a manly decision. I would no longer chafe in the small confines of Sydney-town.

PART TWO

A NEW CALL TO BATTLE

My stumbling announcement that I desired above all else to go to sea met with such an agonised reception from my mother that my spirits were much cast down. But it was my father who, although clearly not welcoming the proposal, finally achieved her grudging acceptance and I signed on board a sealer working out of Bass Strait.

My ardour, however, was soon cooled by my experiences at the fishery. I cannot romanticise these events. War has an heroic dimension and a certain glamour, but no lustre, however superficial, could elevate such deeds to warrant the word 'hunting' being used to describe our cruel slaughter.

We would normally divide into two parties where a seal colony was known to congregate. Upon a given signal both groups would simultaneously descend on the seals and commence the unsparing work of slaying as fast as possible. Here a slight blow on the head was sufficient for the young pups, although it was not quite so simple a task with the adult seals.

It was, also, all too easy to fall on the wet rocks which were made more dangerous by being the living quarters of such a colony; one alone was estimated to contain twenty thousand seals. If a man should happen to fall he could be torn to pieces by a huge adult whose mouth was as large as that of a lion or a mastiff. And, indeed, their ferocious barking and savagery when aroused was reminiscent of a dangerous pack of wild dogs. The roaring of the adult seals, maddened to desperation, and the yelping of the pups, together with the shouts of the crew, produced an appalling pandemonium from which I should have been exceedingly relieved to escape. The work of death was soon over. So too was my enthusiasm for this vocation.

Sometimes it was also the practice to divide our sealing parties into teams of two men each, which teams would then disembark from their ship into smaller boats suitable for hunting excursions. This enabled us to approach the rocky outcrops of Bass Strait which were the barren haunts for large colonies of seals. On one such foray my partner and I made for just such a rock. In the absence of any facility for securing our small boat to this rock with safety, we moored it to some kelp at a short distance offshore. We then took our clubs and knives and easily swam to the outcrop which tempted us with its teeming population of seals. We were so absorbed in our bloody slaughter that we were taken unawares by an impending storm.

As we made all haste to gather our skins together preparatory to swimming to the boat, we observed to our horror two enormous sharks. These creatures were gliding stealthily to and fro between the rock and our boat and had doubtless been attracted by the blood from our slaughter. To remain on the rock was impossible because it would soon be swept by the sea. Our predicament was extreme, until we resolved to feed the sharks our accumulated catch. When it seemed that even these ferocious appetites had been satisfied, and the creatures had left behind our latest offering and glided silently away, then, and only then, did we dare attempt an escape. With a desperate plunge into the water we struck out wildly for the boat which we reached without molestation.

Mishaps like this were accounted an everyday part of sealing but they awoke in me a determination to find another seafaring vocation. And it was not long before my youthful ardour was transferred to another maritime activity. Often in these sealing waters we caught sight of a whale's unmistakable spout and the proud display of his mighty tail uplifted against the sky. One of my sealing companions had served on a whaleship and I could not hear enough of his exploits. These heroic accounts differed so greatly from our own seal hunting that I resolved to sign up on board a whaler, for even the smallest lay, as soon as ever I could.

Whaling was hardly the mindless slaughter it had been with the seals. There we simply swam to rocky outcrops or rowed ashore and proceeded with our dull-witted killing. Wading through a colony of seals and bludgeoning them to death, skinning them on the spot and heaving our prize back to the boat could never compare with hunting the whale. I know that now, thirty years later, there are some few voices being raised against this killing.They point to the diminishing annual catch and compare it to the seal trade which was completely worked out by the mid-1830s. But these stay-at-home zealots have never faced the whale or else they would have realised that this was not a one-sided confrontation.

I have seen stove boats and crushed and maimed sailors caught by the thrashing flukes of an adversary who then quit the battle and swam on like some stately galleon of old. And so I say to you, be economical with your lamps and candles. Not a gallon of whale oil you burn but men's lives have paid for it in their battle with Leviathan. And it was this battle which made a fundamental impression on my young and eager mind; an impression which has never left me to this day.

When I was first within earshot of the whale's massive tail as it struck the water or splintered a daring whaleboat, it instantly took me back to the sound of cannon fire at the Battle of Algiers. This thunderous concussion resounding across the water was the angry sound of a new kind of battle. But this was not all. It was not an idle fancy to imagine that the wreath of vapour from the whale's spiracle was even akin to the smoke from the touch-hole of a cannon.

It is also a fact that the quintessential business of whaling is the same as that of battle — slaughter. Except for soldiers, whalers have to look danger more fully and steadily in the face than any other class of man. And the awful truth is that young sailors, like young soldiers, make the best killers. Most whaleship owners sit in their counting-houses like generals in their war rooms and send us young men out to do their bloody business.

Eventually, if victory was to be ours, the whale would signal its defeat by spouting a red flag, a pink mist punctuated with barrel-sized gouts of blood and gore. Our boats would then beat a hasty retreat to avoid the dangerous thrashing as the whale drowned amid a flurry of its own blood. And then it would float alongside like a conquered fortress, with our flag of capture displayed on a whale-pole jutting from the top of the vanquished foe. This was how the battle was won.

Even the most peaceable of men can find their ardour inflamed at the prospect of a battle with this mighty adversary. I have met pacifist Yankee whaling captains who were zealous in keeping their ships free of swearing and profanity and in leading their crew in prayers; who regarded card-playing as a sin and who would not do battle with whales on a Sunday. And yet these same New England Quakers made enviable whaling captains. Their peace-loving creed met its only exception in the stern war they waged against Leviathan.

And truly, the whale is a worthy adversary when pitted against man in the telling battle of strength and will and wiles. Furthermore, I soon came to see, and still believe, that our whaling captains were mighty explorers. They navigated uncharted seas and mapped archipelagos where Cook had never sailed. If men-o'-war now peacefully ride in once savage harbours, let them fire salutes to the honour and the glory of the whaleships whose anonymous captains originally showed them the way. All that is made such a flourish of in the old South Sea voyages, those things were the lifetime commonplaces of our whaling captains.

Often, indeed, adventures to which Cook, Flinders and their company dedicate three chapters in their published accounts have been accounted unworthy to set down in whaling logs. Remember the despondent tone adopted by our celebrated Captain Cook when he described the beauty and fertility of the islands of the Pacific Ocean and lamented, in hapless sorrow, that no charm of lucre existed upon their shores to induce civilised nations to cull their produce. How little could he have anticipated that, in a few years, numerous British and American and now even Australian fleets would be constantly traversing that ocean, engaged in the lucrative commerce of whaling.

My first berth on a whaling ship was in 1821. Here I signed on as a greenhand aboard the *Jupiter*. Seamen are notoriously careless about the nature of the articles they sign for a sojourn at sea, their chief anxiety being to get underway. And so it was on this voyage. I found that many of my companions had signed and engaged themselves to go to places unknown and were to be paid an uncertain amount for a long voyage of undetermined duration. On our first morning out, when we were called to muster, I expected an address from the Captain not unlike those I had been accustomed to on the *Leander*. My expectations were soon dashed. This was a much more makeshift affair and the Captain hardly more educated than his motley crew.

To be fair, though, I could see that we were a mixed crew of greenhands and experienced whalers who were now to be thrown together at sea for upwards of five months. It needed a strong hand and a gruff voice to establish discipline from the very start. It is sometimes supposed that but very little discipline is observed at sea, except aboard a man-o'-war. This, however, is quite wrong and although our ship may make no higher pretensions than those of a whaler, yet obedience is as well understood here as on the decks of a line-of-battle ship. It is necessary that such should be the case for, in all stations of command over others, the forms of respect must be rigidly adhered to or all authority ceases.

Our master's preamble certainly made such distinctions clear. 'I suppose you all know what you came whaling for?' he began. 'If you don't, I'll tell you. You came to make a voyage and I intend you shall make one. You came for oil. You came to work. And don't forget that I'm here to see you work. By God, you'll earn your lay.'

Here he took a turn on the quarterdeck, concentrating his ideas for another burst of eloquence. 'Well, boys, this will be a greasy voyage, a monstrous greasy one if we are to fill all the barrels and our pockets besides. You will do as the officers tell you and work when there's work to be done. We didn't ship you to be idle here. And another thing — I allow no fighting aboard this ship. Come aft to me when you have any quarrels and I'll settle 'em. I'll do the quarrelling for you, I will. So, if there's any fighting to be done, I want to have a hand in it. And I'll have no swearing neither. I don't want to hear nobody swear. This is an infernal bad practice, it breeds ill will. If I catch any one at it, damn me, I'll flog him, that's all.' This was followed by a nod of the head, as much as to say that he meant to be as good as his word, which we mustn't doubt.

'When it's your turn below you can stay there or go for'ard, as you please. But when it's your watch on deck you will be there on time and you'll work. I won't have no skulking. Any of you that I catch below when it's your turn on deck, except in cases of sickness, shall stay at work until I think it proper to relieve you.' A stride or two aft and a glance to windward confirmed his resolve. 'You shall have enough vittles to eat, plenty of it if you work. But if you don't work, you'll starve. And it won't do to grumble about the food.' A mysterious shaking of his head implied a vast deal of meaning and a considerable threat. 'If you don't get enough, come aft and apply to me. I'm the man to apply to — I'm the Captain.' And here he surveyed us all with a look which seemed to say not only that he was the Captain but that it was a source of infinite satisfaction to him that he was.

'Now, the sooner you get a cargo of oil on board, the sooner you'll get back home. Do your duty, act well your part towards me and the officers, and we'll treat you well. But if you show any obstinacy or cut up then I'll be damned if it won't be the worse for you!' This was followed by three or four strides about the quarterdeck and a portentous silence for some minutes, as if he was deliberating on some grand finale. But inspiration failed to descend and we were dismissed with a gruff: 'That's all now. Go for'ard where you belong.'

PITY THE POOR GREENHAND

The deck of any whaleship about to weigh anchor for an extended cruise presents a scene of confusion and disorder. Our first task was to clear the decks. Where space was seriously limited, every single piece of gear had to be allocated its special place. This meant that the mates could produce what was wanted at a moment's notice. Then everything was checked to ensure that all necessary items were aboard and that all were in sound condition: knives, hatchets, drogues, tubs, lugsail and boom, boathooks, harpoons, lances and line and water butts. And then only incidentally, or so it seemed to me, was a passing observation given to our provisions.

Among the few experienced members of our crew, many had come aboard only a few hours before and had arrived in a drunken stupor. Some had literally been dumped into the fo'c's'le where they lay grog-soaked and unmoving. While some thus resisted all efforts to rouse them out upon deck, others were able to respond clumsily to the familiar orders and imprecations of the mates.

With us greenhands, however, although we displayed a serious sobriety we were even more helpless because of our ignorance of anything pertaining to the intricate world of spars and rigging. Faced with these twin handicaps of inebrity and inexperience, the mates were forced to undertake the routine operations of getting the vessel underway with the small handful of men who were serviceable, at times assisting in the execution of their own orders. Meanwhile, they kept up a succession of admonitions, curses and blows, calculated to elicit as much activity as possible.

It was most unusual for a whaleship to sail with the same crew from one voyage to the next. But at least I had one considerable advantage in having spent time at sea before. Unlike my totally inexperienced companions I was not paralysed with seasickness when, after several hours of this pandemonium, the vessel finally struck boldly out to sea. Then these poor fellows received only the dubious offer of whaling's traditional remedy. The cook brought a piece of pork fat about as big as a walnut and tied on to a string. His advice was to coat this with molasses, swallow it, pull it out by the string and repeat this nauseating process two or three times. There were even some who were sick enough to try this desperate remedy!

It was no consolation to the men to be told that the great Horatio Nelson had been equally prone to constant seasickness. Clumsy and painful though their efforts might be, they were forced to go though the form of taking part in every task demanded of the crew. If they pleaded illness or fear, a curse or a blow from one of the mates served as an irresistible argument. And certainly no incentive less violent and menacing would have succeeded, for few situations are less likely to encourage voluntary activity. Eventually, however, the ridicule of seasoned shipmates and the perils of whaling would underscore the necessity of learning the business quickly.

In a whaleship may be found men of all classes, for the business is, in fact, a meeting place for every kind of adventurer on the ocean. Often they are men reduced in their circumstances by intemperance or misfortune, who, in a fit of despondency, have entered on board for a whaling voyage. They often have no specific object in view but only a vague idea and are kept going along in a kind of delusion until their ship sails. Then, when the vast ocean separates them from family and friends, they arouse themselves to the recollection of what and where they are and what and where they might have been. It is very rare for greenhands to recover anything commensurate with their time and labour. However, if they continue in the business and rise to harpooner, officer or even captain then they will become men of wealth and standing in society.

I will also admit that the task of the mates, too, is no light or pleasant one. Starting out with a crew whose members, with few exceptions, are intoxicated, seasick or inexperienced, they face the necessity of creating and preserving a spirit of discipline, of familiarising the men with manifold duties and of building up a state of collective efficiency. Inebrity and seasickness soon wear away, but the transformation of inexperience and callous indifference into effective labour is a long, slow process accompanied by the ruthless and often brutal enforcement of a stern policy.

Official steps in this process were taken on the first day out, when the whole crew was called aft to be divided into watches. The two watches were headed usually by the first and second mates and those officers made alternate choices until all our names had been drawn. There are larboard and starboard watches which keep vigil upon deck for four hours at a time. Then our boat crews were organised and the Captain and each of the mates, in order of rank, chose the hands who were to man their respective whaleboats. Each of our three boats would be manned by a crew of six, including the officer in command.

Those with the most expertise were accorded the envied role of harpooner, while the strongest seamen were chosen as pulling hands on the boats. It was our Captain's task to weld us into efficient whaleboat crews as quickly as possible and practice runs became our daily routine. Only the cook, the steward and some few others would remain behind and not participate in the whale hunt. These were the shipkeepers whose duty it was to handle the ship while we hunted from the whaleboats. One minute we would be going about our allotted shipboard duties and the next a cry was raised and everyone raced for the boats. All was 'hoist and lower' until we were adept enough to do it in the dark. In the little universe that our ship became, each sailor had to learn his alloted tasks in a highly rigid hierarchy.

My experience aboard the *Leander* had taught me that naval custom had not sprung into being arbitrarily. The reason that a chain of command located every crew member so precisely was mortally important. In emergencies, every man had to know what to do and the only way he could know this was by being told by someone with more experience and knowledge higher up the chain. Frequently, the safety of the entire ship depended not only on the order coming down in a clear and timely manner, but also on its being executed immediately and without question.

And here on our whaleship such proceedures were equally effective. Thus, when the Captain ordered a change in course, the helmsman was required to repeat the order aloud, precisely as given, to ensure that he had heard it correctly. Failure to do so could compromise the very safety of the ship. I found once again that, as it had been in the Navy, careful training under the threat of danger bred a unity in men.

Here too I learned that there were certain forms of respect which were never deviated from. As foremost hands we never went aft unless we had business which called us there. And then we would take the lee side of the ship and any skylarking on the quarterdeck would be severely punished. Furthermore, if one of us had occasion to go into the inner sanctum of the Captain's cabin, that man would be careful to leave his hat out on deck.

After the crews for the whaleboats had been chosen, men were sent up the mast for the first lookout and our voyage was officially deemed to have started. Because I was one of the youngest in the crew and nimble aloft, I was chosen to help man the masthead. This is the least arduous task on a whaleship but it does require keen eyes. The lynx-eyed watcher sends forth his anxious regards upon the ocean and deems his station a post of honour should he be the first to discover a whale within pursuing distance. Here my training and experience on the *Leander* were soon recognised as I clambered aloft with a sure foot and an eager hand.

Those who had kept to their berths for the last few days declared that even to look up at the masts swaying with the motion of the ship made them deathly sick again. 'As for climbing up there,' said one with unfounded belief, 'that is quite impossible and I will not be fool enough to try it.'

With what dismay, therefore, did they hear on the third day out, as word passed below, that all greenhands were to come on deck to practise running up the rigging. With doleful groans and dolorous countenances they most solemnly asserted the utter impossibility of such an undertaking and the certainty of their falling before they got six feet above deck. Paler than so many ghosts, they were forced to mount the rigging where they took the shrouds in a close embrace and directed looks of terror and dismay down at the mate. How great was their relief when they once more found themselves back on deck!

There is nothing the inexperienced on board find so difficult to grow accustomed to as the difference in rank and consequent difference in physical comforts which prevail on board ship. Why must the Captain and his three mates have more space allotted to them, or upon what principle of right or justice do the officers dine upon supposed delicacies while foremast Jack soaks his hard biscuit in a decoction of oak leaves sweetened with molasses which goes by the name of tea? How, they ask, under a democratic flag, can the

Captain order them off the quarterdeck, the most pleasant part of the vessel, and point to the wretched hole forward of the windlass as their appropriate sphere? All this and much else the poor greenhand, brought up in the belief that one man is as good as another, can never properly understand although he is obliged to submit.

In our first days at sea, yards were braced and sails were reefed, furled and shaken out to create the necessary tension above. Then, as the weather worked upon the rigging, it loosened and stretched and required our constant maintenance. With over twenty sails, and each rope identified by name, there was much to bewilder a newcomer and at first the labyrinthine confusion of spars and ropes was as alien as a gigantic spider's web. And underscoring all of these lessons with its ominous music, these first few days were filled with the sound of hasps honing the steel of our lances, cutting spades and harpoons. Our deck was alive with blinding flashes from the overhead sun and as we sailed on, our killing tools grew sharper and sharper.

All the crew were now to some degree becoming broken in to life at sea and pretty soon the greenhands were even claiming for themselves great credit for their seamanship. With infinite pains they had been taught a few of the many splices, knots and ties which all old tars have at their fingers' ends. By dint of steady perseverance they were now able to chew tobacco without being nauseated and to spit about the decks without feeling that they were committing a dirty trick. They could hitch up their suspenderless trousers and cock their hats in a manner faintly resembling that supposed to be peculiar to the genuine tar. And so they called themselves seamen. They had arrived at the summit of their tree of knowledge and did not fail to congratulate themselves upon the prospect stretching out before them.

Alas! They still had one final mortification to undergo, one more difficulty to overcome, yet another branch of the business to familiarise themselves with before they could call themselves whalemen. What old salt does not know that there is as much difference between a novice and a true sailor as there is between a child's tin trumpet and the bugle which calls men to battle?

Shortly after breakfast, word was passed to man the boats. We were to take some practice in pulling and manoeuvring and preparing ourselves for our ultimate purpose afloat: the chase and capture of the whale. It being a calm day, the crews were directed to follow the boats down; that is to say, to slide down by the side of the ship abreast of their respective boats in readiness to jump in as soon as the craft touched the water. In endeavouring to jump into his boat — the ship just at that moment giving an unexpected lurch — one of our too optimistic newcomers dropped into the water. All this provided great amusement to the onlookers.

But this was only the beginning of the day's sport. Although the sea wore a smooth face, there was enough of a groundswell to make the use of oars a matter of some difficulty to those who were now holding them for the first time. Having somehow formed the whaleboats into a line, the first mate proposed a race. But this soon discovered the crews' lack of skill. First they had to be taught to keep stroke — to place their oars in the water all at the same time and to lift them out again with one motion.

While the necessity of this skill was perfectly understood, still it was no easy task to achieve. As the swell lifted the boat, those not paying strict attention would fail to reach the water with the blades of their oars. When they didn't meet the resistance upon which they'd counted, they would invariably tumble over on their backs, heels high in mid-air, heads under the seats and oars dangling askew.

My exploits and example aloft were much more commendable. Here I would reach a pair of cross-trees fitted to the mast. They provided the base for a tiny platform on each side of the mast; above this foothold was suspended a couple of padded hoops. No matter how vigorously the ship plunged beneath, these hoops would hold the lookout secure and a sharp-eyed fellow in this embrace could spot a whale's spout five or six miles away. But when I descended from my lofty perch, my eyes often ached and burned for an hour afterwards from the intensity of gazing across the shining expanse of waters.

I soon learned that the ultimate guarantee of a ship's safety is the alertness of its lookouts. When close to land, of course, regular soundings are taken with lead and line, but a good seaman has a restless eye — one which is always shifting between the weather side of his boat, the masthead, the spars and rigging above, the sea around and the horizon ahead. And so I became aware of the most minor and, to a landsman, insignificant changes in the sea and the sky. In this way I could detect an early warning of a change in wind or weather. Even a change in the pattern of the clouds or an unusual tint to the sky became enough to alert me.

I was an eager and willing pupil and some of my lessons were simple enough. Close to land, the dark colour of the sea became paler and fish were usually more numerous. Seaweed, driftwood and other flotsam often indicated off-shore currents. The on- and off-shore breezes and the flight of birds that mimicked them — outward at dawn and returning at dusk — could also be detected far from land. Finally, the scents of land might be carried far out to sea on a breeze — the unmistakable smell of earth after rain, of wood smoke and of swamps. And then close to home there was the pungent and always welcome smell of eucalypts.

Sharp eyes and a keen desire to acquit myself creditably in front of our gruff Captain kept me alert and ever vigilant while aloft, although others in this solitary roost can succumb to dreamy reverie if not to actual sleep. There you stand, surveying a watery domain, while beneath your ship, in the ocean's unfathomable depths, swim huge monsters of the sea. This was what my experience as an infant atop the eyrie at home had epitomised. And besides, I was soon accounted a lucky choice, for it was my good fortune to spy out the first whale we took. Truly, I had now found myself a role and a future that seemed to have been marked out for me since birth.

FO'C'S'LE YARNS

The hazardous life of the experienced whaler will always be a story of extreme adventures. Narrow escapes from death mark his career and the ordinary dangers to which he is exposed often blanch the cheek of the novice. After supper on any whaleship, from six to eight o'clock, was the season for amusements of every variety. The officers would be conversing together in the after part of the ship, while the men would assemble around the windlass and smoke their pipes. It was a time well suited to reverie and yarning.

Three types of persons were always in great demand at such times: the man who could play a musical instrument; a 'minstrel boy' who possessed a wide repertoire of songs and a passable voice; and the seaman who could spin a good yarn. All three would always be heard with interest and respect and there was a joyous contagion about boisterous songs for which the leader produced both familiar and extemporised verses while all of us roared out the choruses. In particular, though, a good yarn, well told, had a never-ending appeal. The tedious hours of the dogwatch in the evening were often beguiled away by the men with their tales of adventures during former voyages.

On a night early into our voyage, we were gathered about the fo'c's'le scuttle and the question of whale-fishing came up for discussion. Until that time, strange as it may seem, no word of this had been mooted. Every man seemed to shun the subject, as if he were afraid of frightening off our prize or, in the whaling parlance that I was learning daily, 'gallying' the fish. Once the question was voiced, however, nearly all had something to say about it and many addleheaded opinions were ventilated. It was clear that among so many greenhands there was little, if any, genuine knowledge. The knowing ones were wise about whales and we learned much of fo'c's'le natural history in these evening recitals.

One of our companions, a rare example of an experienced whaler in the fo'c's'le, entertained us with his narration of a miraculous survival. He told of an earlier voyage when he was at his post in the bow of a whaleboat, tending to the line, when that most dreaded of all accidents occurred — he became entangled with it and was rushed overboard. At once the boatsteerer seized the hatchet, which is always at hand, and chopped the line with the faint hope that when it slackened his crewmate could extricate himself.

The accident being so sudden and dreadful as almost to stupefy the crew, none of them spoke a word, but each eye was fixed upon the sea with fearful interest. Several minutes elapsed and the last hope was expiring when an object was seen to rise to the surface a short way from the boat. 'They told me that I showed no sign of animation and was apparently lifeless as they lifted me into the boat. But since you see me before you this day, alive and well, I can tell you of my descent beneath the waves.'

By now, of course, we were all an eager audience, as enmeshed in his yarn as he had been in the tangled line. We urged him to continue. 'I'll bet that none of you can equal my yarn,' he boasted, displaying an enviable plug of tobacco.

'One minute I had been stooping down to clear the line which had caught in the bow. As I attempted to throw it out from the chock so that it might run free, I was caught by a turn around my left wrist. The next minute I was being rushed down into the green depths with unknown force and swiftness. I feared that my arm would be torn from my body, so great was the force of the water during my descent. I knew well enough that my only chance of life was to cut the line but it was impossible to move my other arm from my side. It was pressed there by the sheer force of the element through which I was being dragged.

'I determined that if this was death, I would face it with open eyes. It appeared as if a stream of fire was passing before me which darkened as I descended. I remember feeling a terrible pressure on my brain and a roaring as of thunder in my ears. And all the while I was horribly conscious of my situation and still made efforts to reach the knife secured in my belt. At last, when I knew that my strength was failing and my brain reeling, the line for an instant slackened, giving me a chance to grasp my knife. In the instant that the line again became taut I set the blade to its edge. After this I remember a feeling of intense suffocation, a gurgling spasm, and then nothing until I awoke to an agonising sense of pain, back in the whaleboat.'

And so our companion was living proof that death could be cheated and that night he became an object of reverential awe among us.

Some little time later a hitherto self-effacing sailor, the only one who had seemed unimpressed by this descent into the watery depths, finally added his contribution to our evening sessions. 'Come to anchor out of the cold and I'll tell you a real adventure,' he boasted, casting a critical eye upon our current hero.

On one occasion, he told us, his boat had been among a school of whales, when another boat close by was placed in a very critical position. A large whale that had been struck suddenly raised its flukes up above the crew as if ready to annihilate them in an instant. The terrified men sprang out of the boat to avert the impending disaster

— all except this one who retained his post undaunted, awaiting the fury of the monster. The whale, however, withdrew his flukes and by this action gently landed two of the crew, who were hanging on to the gunwale, back into their boat!

'I assure you,' our narrator declared, 'that I would not venture to state a thing apparently so incredible had I not been witness to it myself.' Truly this yarn eclipsed all other attempts during our evening vigils in the fo'c's'le. Now, however, I know that our storyteller was entertaining himself with our astonished reactions, for I have since heard this yarn circulated to dupe many a credulous greenhand and always told with equal assurances.

I was appealed to constantly to support this or that facet in these stories but, like most of the others, could only draw upon my imagination. How did a whale act, what chance was there of being saved if your boat got smashed, and so on into infinity. And truly we needed these distractions, for conditions on board were often unpleasant and sometimes hardly to be endured.

The monotony of our diet was one of the first things that had to be accommodated. Salt meat, hardtack, beans and rice, and biscuit as hard as a pantile were our staple fare. It was surely in revolt at this unrelenting offering that some of the crew began to indulge in other 'delicacies'. The novel delight of fritters of whale brains mixed with flour and fried in whale oil seemed of dubious appeal. But even less tempting was an unholy concoction of stewed barnacles, which were said to be especially tasty when taken from the blowhole of the whale!

These experiences, at least, could be avoided by those with more fastidious stomachs. Less easily repulsed was the nightly attack from cockroaches which descended on us in unrelenting droves. These scuttling armies made a noise like a flush of quail among dry leaves. Often on a warm Pacific night a sudden curse from one of my companions would rend the silence as he awoke to find these scavengers nibbling around his lips. The creatures have teeth so sharp that it is said they can eat off the edge of a razor! Whatever their abilities, this I do know from personal experience — I even found one of them up in the main-top, wandering on high very much at his leisure.

Sailors' yarns are not often considered to be amusing anecdotes. The privations we endure and the dangers we confront daily supply us more readily with a repertoire of stories much less unlikely to lift the spirits. They seem inevitably to encourage an even more sombre note in other crew members.

No single yarn exchanged in the fo'c's'le ever affected me more than one from an old seaman. Indeed, its effect on the whole audience was such that it quite gallied us all by awakening our most dire apprehensions and superstitious dread.

Many years earlier this shipmate had been sailing not far from Cape Horn. The conflict of the elements was a sight that could not but inspire the beholder with awe. Frequent squalls of snow and sleet swept across the ship while mighty surges, swollen with accumulated water, gathered in fearsome array at their rear. Then these advancing waves swept onward in pursuit with increasing momentum until they towered above and threatened to engulf the vessel.

'Iceberg in sight!' suddenly echoed through the ship and everyone hurried on deck to view this interesting but unwelcome visitor. Because their position can never be foreseen by calculation, as that of rocks laid down upon our charts, icebergs are objects of peculiar dread to the mariner. Through the gloom of the driving snowstorm, its craggy peaks of light green were distinctly seen rising above the angry waters. It had a wild and dreary aspect which comported well with the hoarse cries of the wind and angry sea.

Everyone was reminded of the forlorn regions to the south. Many vessels bound around Cape Horn, that have never been heard from again, are supposed to have been lost among these ice islands.

Then at dawn, the wind rose to a gale and a succession of thundering, grinding noises gave fearful evidence that all around in the thickly forming fog there were other icebergs. In the morning, when the winds abated, the Captain found to his great joy that the ship had not sustained any serious injury. It was remarked with surprise that the accumulated icebergs had been separated by the wind and that in one place a canal of open sea wound its course among them as far as the eye could discern.

It was two miles beyond the entrance of this canal that a stranger made her appearance about noon. At first it seemed likely that there would be a friendly 'gam' with this unexpected newcomer, but as they drew near everyone was struck with the strange manner in which her sails were disposed and with the dismantled aspect of her yards and rigging. It was no surprise when she grounded upon some low icebergs and at once the Captain took a boat and rowed towards her.

On approach it was observed that her hull was miserably weatherbeaten. Not a soul appeared on deck and, although they hailed several times, no answer was returned. Already there was a feeling of great disquiet and this unease only increased when the Captain determined to set foot on board. Before this could be achieved his eye was caught by an open porthole. On looking into it he could very dimly discern a man reclining in a chair. No amount of noise, however, could rouse this figure, so the Captain continued in his resolve to step aboard. Once there, he removed the hatchway and descended into the cabin with his men.

A tremor seized him when he entered, for here he found that the inmate was insensible to strangers. The Captain was faced with a corpse; a green damp mould covered the dead man's cheeks and forehead and veiled his eyeballs. He had a pen in his hand and a logbook lay before him whose last entry was dated almost a dozen years earlier. It read: 'We have now been inclosed in the ice for seventeen days. Our fire went out yesterday and our master has been trying ever since to kindle it again without success. His young son died this morning. God have mercy on us if there is no relief soon.'

Everyone hurried from the room with scarcely a word. Nearby, in the principal cabin, they found the frozen body of a woman stretched out on her bed. Next to her, on the floor, was the corpse of a man. In one hand he held a flint and had a steel in the other, as if in the act of striking fire upon some tinder scattered around. In the forepart of the vessel they found several sailors lying dead in their berths and the body of a boy was crouched at the bottom of the gangway stairs. The whole scene was harrowing in the extreme.

Neither provisions nor fuel could be discovered anywhere. Eventually, the Captain was discouraged from examining the death ship still further by his men's growing dread. He therefore took the logbook and sailed off, with everyone deeply affected by the awful example.

At last, getting tired of these stories in the fo'c's'le, I went aft for a drink of water before turning in. The harpooners and officers were grouped in the waist, earnestly discussing the very same subject, but their talk at least carried the ballast of actual experience. As I passed one harpooner I heard him say: 'Feels like whale about. I bet a plug of tobacco we raise whales tomorrow.' Nobody took his bet; it appeared that they were mostly of the same mind.

A BEACON AT NIGHT

On the very next afternoon, after a morning which had been rather stormy, I was at my post on lookout when a school of young whales made their appearance close to the ship. 'There she blows! There! There! There! She blows! She blows!' I yelled. 'Where away?' came the answering cry. 'There go flukes, on the lee-beam, about two miles off. A school of whales,' I called.

At sea, with no land in sight, no bird or other living thing, in the cold monotonous grey of early morning, imagine a lone whale steering in solitary grandeur, his path marked at times by the cloud of his breath. All this has something sublime about it. You may have seen but a glimpse, the smallest portion of that great mass — a black line, scarcely distinguishable from the deep blue of the sea.

Now multiply that and you will have some idea of the sight of the pod of whales we had encountered. They drew near with a scarcely perceptible gliding motion, generally rising in pairs and showing a tremendous length of back. As their blowholes came up to the surface there arose a jet of vapour which spread out into a cloud and then melted away. Our ship seemed in no way to disturb them.

Now and then they would rise quite close to the side, when we would distinctly hear a whistling sound during their spouting.

No sooner had my excited cries rung out than our Captain ordered the mates to lower their boats, while he did the same with his own. While the direction and size of the pod were being reported, the off-duty hands rushed out of the fo'c's'le companionway, rolling up their cuffs, tightening their belts and peering over the rail for a clear sight of the quarry. We managed to get near the whales but were then unfortunately sighted before we could get close enough to dart any harpoons. The consequence was that the pod separated and went off with great swiftness in different directions.

One whale, however, after making several turns came at length right towards the Captain's boat, which observed and awaited its prey without moving an oar. When the young bull came close by and received a blow from a harpoon, it appeared only numb for a few seconds. Then, suddenly, it recovered itself and darted off with such speed that when the Captain's boat spun round it was within a miracle of being upset.

Away they went, dead to windward, at the rate of twelve or fifteen miles an hour, right against a head sea. The water flew over the bows of the boat with uncommon force, so that at times it appeared to be ploughing through it, making a high bank of surf on each side. The second mate, having observed this action, managed to waylay them. A short warp was thrown and both boats were soon being towed along, without the speed of the whale appearing in the least diminished. The other two boats finally regained the whaleship.

In an alarmingly short time the two crews fast to the whale had disappeared altogether, being at too great a distance to be seen with the naked eye. With the aid of a telescope I could just discern them like mere specks upon the ocean. I watched them towed through a good deal of white water until it was no longer possible to trace anything, even in the most indistinct manner. When I alerted the crew below to this disappearance, they took a bearing by compass of the direction in which I had last seen them. In this way we might continue to beat the ship up to that quarter.

It was now within half an hour of sunset and there was every appearance of the coming on of an ugly night, as a seaman would say. I remained aloft until I saw the sun dip below the troubled horizon. The winds then increased with heavy squalls at times so that we were obliged to double-reef our topsails. And now our painful and precarious situation bore most heavily upon us. Our Captain and second mate with ten of the crew had disappeared and we were far from land. We continued to burn blue lights and we had likewise a tank containing oil and unravelled rope burning over the stern rail as a beacon for our lost companions. But although all eyes were employed in every direction, no vestige of them could be seen.

The wind howled through the rigging, the waves beat against our sides. I had taken my place again on the masthead and once I even imagined that I heard the Captain's voice ordering the ship to bear up. Others, in their anxiety, swore they had sighted the boats, only to be mistaken. They had strained their eyes though the gloom until fancy robbed them of their true speculation and left her phantasmagoria in exchange. There were not many on board who did not think of home on that dreadful night; there were not many among us who did not curse this sea and all sea-going avocations. At the same time, and with the same breath, we all blessed the safe and cheerful fireside of parents and friends back home, who at that moment we would have given all we possessed but to be with.

Then, at the very moment when despair had firmly settled upon us, I discerned a light ahead of the ship — just as we were going about and by that action could easily have missed it. Everyone looked eagerly in the direction I gave and in a few minutes it could be plainly perceived. In a short time we were close up to it; then, to our great joy, we discovered the Captain and all the men, lying to leeward of the dead whale. Its enormous and sheltering bulk had in some measure saved them from the violence of the storm. We secured the prize alongside, although there was every expectation that it would be lost in the night from the roughness of the weather, and then all were safely delivered aboard.

A peculiarity of all these vessels, as I afterwards learned, was that no stated allowance of anything was made. Even the very water was not served out to us, but was kept in a great scuttle-butt by the cabin door to which everyone who needed a drink had to go and from which none might be carried away. No water was allowed for washing except from the sea, and everyone knows, or should know, that neither flesh nor clothes can be cleansed with that. But a cask with a perforated top was lashed by the bowsprit and was kept filled with urine, which I was solemnly assured was the finest dirt-extractor in the world for clothes. The mates, however, even with this recommendation, did not avail themselves of its virtues. They were content to use ley, which was furnished in plenty by the ashes from our galley fire. Of course, when rain fell we had a good wash — that is, if it was night and no other work was toward. Yet we were not allowed to store any rainwater for washing purposes.

Another curious but absolutely necessary custom prevailed in consequence of the short commons under which we lived. When the portion of meat was brought down in its wooden tub at dinner time, it was duly divided as fairly as possible into as many parts as there were mouths. Then one man turned his back upon the carver, who, holding up each portion, called out: 'Who's this for?' Whatever name was mentioned by the arbitrator, that man owning it received the piece and had perforce to be satisfied therewith. Thus justice was done to all in the best way possible and we were saved any friction whatsoever.

As some of us were without clothes except what we stood upright in when we joined, the 'slop chest' was opened and each applicant received from the steward what our Captain thought fit to let him have, being debited with the cost against such wages as he might afterwards earn. Soap, matches and most especially tobacco were likewise supplied on the same terms, but at higher prices than I had ever heard of before for these necessaries. After careful enquiry, I ascertained what, in the event of a successful voyage, we were likely to earn. Each of us was on the two-hundredth 'lay' or share at two hundred pounds per tun, which meant that for every two hundred barrels of whale-oil taken on board we were entitled to one at four pounds per barrel. It was now that the magnificence of our outlook at this voyage's end began to appear much less promising.

RIDING THE WHALE'S BACK

None of these initiations, of course, could swerve me from a heartfelt desire to take a turn in a whaleboat myself and I soon achieved this wish. Here I found that if speed and grace of line were missing in our whaleship, then they were possessed in a superlative degree by the whaleboat. Being still a boy I was given the stroke oar in the boat, which is one of the shortest two in that vessel. And just as in any sort of battle there is an art of warfare, I had already learned that whaling was no exception. What might look like a ferocious assault is really a planned campaign. Nowhere is this more evident than in the strategic way we would approach our enemy. Because the whale's eyes are on the sides of its enormous head, its front and rear vision is greatly limited. Head and tail, therefore, are the customary approaches to this battle, with a frontal attack preferred because it avoids the deadly threshing of the whale's broad tail fins, or flukes.

Our other whaleboats were flying along behind us, spreading wider apart as they came, while in the bows of each stood their harpooner with his right hand on his first iron. Four boats were down and heading to leeward, their course divergent, so that at two miles from the ship we peaked our oars with a space of about one-third of a mile between the boats, thus commanding a reach of nearly two miles' front. All of a sudden, at a motion of the mate's hand, the peak of our mainsail was dropped and the boat swung up into the wind, laying hove to and almost stationary. Once our centre-board was lowered to stop her drifting to leeward, my companion alongside explained why we had hauled up.

'Now,' said he, 'the whales have sounded and we must wait till they rise again. Then we'll probably get there or thereabouts before they sound again.' And with this explanation I had to be content. Silently we lay, rocking upon the gentle swell, no other word being spoken by anyone. At the main masthead of our ship was a square blue flag and the ensign at the peak was being dipped. These were signals well understood and promptly acted upon by those in charge of the boats, who were thus guided from a point of view at least a hundred feet above the water. Being at sea level, of course, we had no idea of the whereabouts of our prey.

And yet we were still taken unawares; when the whale breached it sounded ominously close, at our very backs in fact. In truth, though, it was more than a dozen boat-lengths away from us. It surfaced with an almighty hiss and I have to confess that I cried out in fear. This was as much in ignorance as anything else, for like all those pulling on the oars, we could not see our adversary. Instead, we were kept to our task by the steersman, rowing in time to his barked commands and guided by his eyes. We faced the stern and all we knew was that our quarry was close enough to rain down a vapour with an odour like a bank of seaweed. Then a rank oily mist enveloped our boat and the whale was with us. We were urged to be more stalwart with our oars, but it is a terrifying thing to have your back to a monster while every thrust of the oars brings you closer to its awesome bulk.

When the lower jaw of the whale is dropped enough it shows an opening big enough to drive an ox team through, and were the huge animal coming quickly towards a man for the first time, a vision of such horror would be unfolded that he would certainly wish he had not come whaling. We were saved from this horrifying vision; but another terror was yet to come.

Our momentum, together with the sudden breaching of the whale so perilously close, contrived a further test of our already tightened nerves — we went aground on its very back! For one sickening, drawn-out moment I relived John Wilkins' experience in Sydney Harbour in 1790. But what had entertained me in my youth was now a fearful real-life experience. We were momentarily lifted free of the water, our oars all askew, and I am sure that I was not the only one who almost abandoned his post. The desire to turn around and see our foe was irresistible. It seemed preferable to stare death in the face than not to know the visage of this creature we were pitted against.

Yet even in this moment of near-annihilation, I found myself for one surprising moment admiring the gleaming black skin so mountainously reared beneath us. For a mere second my fear was replaced by a fascination with this glossy black silk, glistening as if it were from the finest looms of Europe. I dare say few whalemen respond to the fine romanticism of nature; we are bent on too desperate an errand to be distracted. My chirography would expand into a veritable battalion of capital letters were I to describe the whale's majesty and magnitude to you fairly. But although there may be a perennially fascinating charm in whaling life when viewed from afar, the business truly assumes a grim ugliness at close range. The poetry of the sea has almost always been written by landsmen. It always will be.

The sharp bark of our most urgent orders brought me to my senses. All of us were sternly commanded to secure our posts. By some miracle the boat was returned safely back into the water, upright and sound, and before we even knew how lucky we were, our harpooner was delivering a very different fortune to this rearing devil. He lunged at the whale as if possessed by a destroying demon and his half-articulate yells of rage and blasphemy rose above all other sounds. Our line hissed out, fathoms at a time, and we were again at the mercy of our foe, carried on a grim sleigh ride that threatened to topple us into the sea. The line ran so fast around the loggerhead that the wood was smoking from the friction. Our nearest oarsman splashed water onto it repeatedly to cool it down; but nothing except victory would possibly dampen the heat of our ardour in this battle.

In the exhilaration of this stage of the chase it is fatal to forget a new danger. The line leaps out like a thing with a will of its own and is infinitely treacherous. If it were to catch an arm or leg of one of the distracted crew, as it can easily do, the unfortunate man would be carried out to his death before any assistance could be offered.

But now the voices of our crew were raised in their battle-cry, inspired as much by fear as by excitement, and it produced a fearful commotion. One voice, however, was singularly absent from this din. It has always seemed to me that one cannot help admiring the wonderful contrivance of Nature in constructing the organs of the whale so as to render it incapable of any utterance. Were it favoured with the power of voice, who could resist the dreadful anguish that must be uttered when labouring under such a weight of pain.

Thus our killing thrust was made. It was now up to Providence to see if we remained afloat. I looked desperately at our hatchet in its niche and wished that it was my decision which could sever our line joining us to this malignant beast. He made a fearful to-do, rolling completely over several times backward and forward, at the same time smiting the sea with his tail, making a deafening noise and a mighty pother. He snapped his immense jaws together and thrashed the sea into foam for acres around. I don't think that I even breathed as I watched that awful black mass rising into the sky. You never really know how big a creature it is that you face until you see him eclipsing the sun.

It was impossible to avoid reflecting upon what would happen if, in our unprepared and helpless state, he were, instead of tumbling about in an aimless blind sort of fury, to rush at our boat and try to destroy it. Very few indeed would survive such an attack, unless the tactics were radically altered. No doubt they would be, for practices grow up in consequence of the circumstances with which they have to deal.

And so we were drawn relentlessly on towards the horizon and into the eye of a storm. Lightning forked ahead inside a roiling mass of black cloud. Truly, I believed, the Beast was dragging us to Hell itself, beyond all salvation. This crashing through a head sea could easily start the planking of our boat, let in the seawater and send us all to the bottom.

The speed at which he went made us leap from crest to crest of the waves with a succession of cracks like pistol-shots. Our whaleboat ploughed deeply, leaving banks of water on each side that overtopped our heads, while the swell of the water behind closed back with an angry rush. The flying spray drenched everyone and prevented us from seeing our quarry, but I fully realised that it was nothing compared with what we should have to endure if the wind freshened more. One of our men was kept bailing the water out which came so freely over the bows, but all the rest hauled with all their might upon the line, hoping to get a little closer to the flying monster.

After what seemed a terribly long chase we found his speed slackening and we redoubled our efforts. Now we were close upon him, abreast of his labouring flukes. Up flew that awful tail, descending with a crash upon the water not two feet from us. 'Out oars! Pull, two! Stern, three!' shouted the mate, and as we obeyed our enemy turned to fight. Then might one see how courage and skill were such mighty factors in the apparently unequal contest.

It has since been my experience that whales are often killed quickly and, up to a certain point, many even appear to take their killing easily. I have seen them rise languidly, all the while bleeding copiously from numerous deep lance wounds while columns of blood ejected from the blowhole show that life is rapidly ebbing. But it is also true that the boats may be fast to a whale for many hours without being able to lance him mortally. And then he can become cunning and will prove to be both difficult and dangerous to approach.

In an instant, he may appear as if electrified with new life and with the possession of his full powers. A wild rush to the surface and the enormous tail will be whirled high in the air and descend in a succession of rapid blows, beating the water into a wide spreading sheet of foam. In a fresh contortion of agony, the huge black head upheaves, with the lower jaw dropping and the mouth open. A craggy beard of barnacles encrusts the snout and upper part of the head or will burrow in groups upon the lips and cluster thickly round the eye — as though trying to put it out — and all this imparts a monstrous character which adds to the demonic appearance of the beast.

His cunning is such that it is impossible to tell in what direction he may rush. The headsman shouts 'Stern all!' hoarsely to his crew to keep them clear of the dying beast. Perhaps a boat will be stove or capsized with its crew struggling in white water among fragments of oars and planks; and all the while flocks of gulls hover and scream incessantly overhead. In the varying play of light the creamy froth resolves into deepest blue or takes varying tints of crimson and purple from the blood with which it mingles. A crimson tide washes against the boat.

All these are scenes surrounding a terrible death, which, under the glare of the tropic sun on the long swell of the South Pacific, are indescribably exciting and extraordinary. On this day, however, our chase and final kill were free from such dramatic possibilities.

The whale's great length made it no easy job for him to turn, while our boat, with two oars a-side and the great leverage at the stern supplied by the nineteen-foot steering oar, circled, backed and darted ahead like a living thing. Eventually our boatheader was able to churn his twelve-foot lance deep into the whale's lungs. As he choked on his own blood he also emptied the contents of his stomach and bowels and I could not help thinking that this messy death seemed singularly degrading for such a mighty adversary. His blood, together with vomited squid and all the stinking contents of his insides, produced a sickening sight and odour. But with the battle almost won, even this was welcome as the smell of victory and success.

Turning upon his side the whale began to move in a circular direction, slowly at first, then faster and faster, until he was rushing at tremendous speed, his great head raised quite out of the water at times, clashing his tremendous jaws. Torrents of blood poured from his spout-hole, accompanied by hoarse breaths through his clogged air passages. In his death throes the whale was spasmodically dilating and contracting his spout-hole with sharp, agonised respirations. By now the thick black stream of his life's blood stained the water to a considerable distance around the scene of the affray.

But this was not all. In his extreme convulsions, the whale also rained blood down upon our boat and its crew and, in the frenzy of the kill, we gloried in this show. At last, gush after gush of clotted gore shot into the frighted air as his monstrous heart burst. Then, in his final death flurry, the whale rolled heavily on his side with a dorsal fin in the air. And lest there be any doubt that we had won the battle, our lance blade was thrust into the whale's eye. He never even quivered.

During this intense conflict I had not noticed what now claimed my attention — several great masses of a white semi-transparent-looking substance floating about, of huge size and irregular shape. In pointing this out the mate supposed it was portions of a giant cuttlefish, but he wasn't sure. I was invited to haul a piece alongside and examine it for myself. Secretly I wondered at the indifference shown by this officer of forty years' experience to such a wonderful fact as appeared to be here presented. A boat-hook soon revealed that it was indeed a massive fragment of cuttlefish, a tentacle as thick as a stout man's body with several sucking-discs upon it. These were each about as large as a saucer and on their inner edge were thickly set with hooks all round the rim, sharp as needles. All this was a mighty revelation to me.

My excited cries at these grisly trophies brought forth a colourful opinion from one of my crewmates. 'I believe,' he confided later, 'that the bottom of the bottomless sea is the home of the giant squid. It lays down there, white as daylight in the eternal dark, its great arms coiled around the foundation pillars of islands. Its thousand other arms, bigger than ships' masts, are gathering up dead whales, drowned sailors and foundered ships.' As fascinated as I was by this yarn, it became an unsettling nightmare that haunted my dreams for some days thereafter.

For the first time it was possible to understand that here was a kind of whale that lived upon creatures as large as himself, if one could judge of their bulk by the sample to hand. But now he rolled in the swell and waves broke over his carcass in a low monotonous surf, intensifying the profound silence that succeeded the tumult of our conflict. We cut a hole in the carcass and, with our spoils of war towing behind, headed back to our ship.

As a red tide poured from the whale's side, the slanting sun which played upon this bloodbath sent a florid reflection into our weary faces. With eyes aglow, we quit the battlefield.

AMID THE FIRES OF HELL

One memorable occurrence, of singular interest to us greenhands but quite unremarkable to experienced whalers, was the sudden population of birds which gathered as news of the dead whale somehow spread. This feathered host provided a snowy funeral cortege which, strange to say, only hovered above the corpse or turned the surface of the water into a plain of snow. With the exception of one peculiar-looking bird, they made no attempt to alight upon the carcass. This bird, however, somewhat like a small albatross and distinguished from his companions by his dirty-grey colour, was alone in boldly taking a precarious place on the whale and at once commenced to tear into the blubber.

It was after the frenzy of the capture, that my further initiation into whaling continued and I soon experienced the most gruesome aftermath of battle. Nothing could be more guaranteed to toughen us greenhands to the act of killing than the task required to render our catch into barrels of oil. Now our ship was turned into a factory. Once again I was forcefully reminded of my days on the *Leander,* when the grisly sights and perils of battle simply had to be accommodated. Eventually we would look on all this with the indifference of a veteran on the field of battle, but truly this cutting-in had none of the romantic cachet of the chase.

But even here, amid the gory task of dismembering, it was impossible to lose feeling for the majesty and the might of our adversary. The whale is, after all, a creature with a brain up to six times the size of that of Man. And how much more impressive, in this all-male world of whaling, was it to discover that our prize had a member as long as we we were tall! Yet given these daunting qualities, it was somehow cruel to find that only a relatively small part of our prize was valued — the head, blubber and bone. By far the greater part, tons and tons of it, was simply dumped overboard to create festering rafts of offal swooped on by birds, fish and sharks.

Now that the business of the voyage was fairly begun by the taking of our first whale, the appearance of the ship and her crew woefully changed for the worse. The decks, which had hitherto been kept scrupulously clean, became covered with oil and it was only by keeping a thick coat of sand scattered over them that we were enabled to get about without slipping. The smoke from the try-works blackened every face, so that the watch on deck resembled a party of coal miners. Each rope exposed to its influence was coated with lamp-black, and our clothing saturated with oil. Even our sails, which had been snowy white, received their share of defilement, for as the men sprang aloft from the try-works with besmeared hands and clothes, they could not furl them without leaving a mark.

Our small world afloat was soon dripping with the loathsome stuff. The pores of our skin seemed filled with it; our hair greasy with its presence and the taste of it always in our mouths. The very biscuit we ate glistened with oil. The knife with which we cut our meat left traces of the abominable blubber which nearly choked me as I swallowed. When it was necessary to work on the lee side of the vessel I was compelled to breathe in the fetid smoke of the scrap fires.

From this smell and taste of blubber — raw, boiling and burning — there was no relief or place of refuge.

Yet, as horrible as it was, the old whalemen delighted in it. This foul smoke was incense to their senses. They would even bake biscuits and doughnuts in the seething oil and portions of the whale's lean meat were deemed a delicacy at their daily dinner. I was finally induced by curiosity to try a piece of cooked whale. The raw meat is of a dark red colour, nearly black, and somewhat resembling very coarse beef. It was minced fine and fried but I could not stomach it.

And then, when the ship had been thoroughly scrubbed, the crew cleansed and all made shipshape and tidy, it would only be to hear that cry again: 'There she blows!' And boats would be lowered and another battle would begin.

With the intense fire of the try-works over a wooden deck for weeks together, and with tarred cordage and canvas above, both of which would burn like tinder, it may seem strange that so few ships take fire. Close attention and untiring vigilance can alone prevent it. If the 'pen' under the try-works, which must be kept full of water, happened to spring a leak in the night without being observed, a short time only would be sufficient to envelop us in flames. It is equally dangerous if this water is allowed to evaporate completely and, with fatal forgetfulness, is not renewed. Then the result is terrible. The planks beneath scorch from the infernal heat of the cauldron above and once they are aflame they will no longer support the weight of the furnace. The whole structure of the try-pots will collapse into the between-decks. Ships have then been engulfed and utterly consumed.

It is also necessary that the pots are kept perfectly clean and that no sediment be permitted to adhere to the bottom, else the oil will be dark and of inferior value. One of us, therefore, had to be constantly employed in stirring the mass, while it was the duty of another to skim out the crackling as fast as it floated to the surface.

This residue then contributed to a terrible economy. Although the furnace consumed huge amounts of fuel, much more than could be stored aboard, only a little of our precious wood was used initially. Thereafter, with an irony that appealed to our peculiar sense of humour, the whale literally rendered itself into oil. But the smell of its burning cracklings was too horrible for description. Suffice it to say that it was as though all the ill odours of the world had been gathered together.

And all this time, while the huge carcass was being hacked from above, it was also being savaged from below. This additional attack was the work of sharks, who were mightily attracted to the scene. They would come right up onto the whale's body with a wave, bite out great pieces of flesh, turn over on their bellies and roll off. Several of them were harpooned and two went off with irons in their backs, which seemed as little to annoy them as a small splinter in the thumb of any builder.

I watched over the side of the ship as the greedy pack of sharks mustered around our whale. To those on deck, they showed the most supreme contempt and would frequently roll on their sides and regard us with their dull staring eyes. They only paused in their voracious feeding when a monster of a shark glided into their midst. In deference to his ponderous set of jaws they withdrew somewhat and allowed him to sail up to the carcass.

He was directly under us and I could have touched his back with my hands. With a sinister sweep of his tail this monster glided round to the flukes. The Captain, observing these antics, exclaimed: 'I've a good mind to send a lance through him and see how his majesty will relish it.' No encouragement was necessary but one eager sailor replied: 'Go to it, Captain, there's no knowing how many of us he has eaten in his time.'

All sailors are uncommonly superstitious about sharks and, with good reason, fear their appearance. Oft-times when a whale had been lashed alongside too late in the day for its cutting-in to be completed before nightfall, the crew would be forced into a most unwelcome intimacy. As they lay in their bunks below, their dreams would be hag-ridden by the thrashing tails and vicious rending only inches from their heads. On the other side of their ship's planking, the adversary they feared even more than the whale itself would be virtually eye to eye with them.

One old sailor affirmed that while he was on an earlier voyage, a shark followed their ship for three days. At the time, there were three hands on the sick list and on the third day one of these men died. He was sewn up in a hammock and thrown overboard. Only then did their dreaded companion and harbinger of death swim away. Another sailor swore that a shark had followed for a full day, at the expiration of which time one of the crew fell from the masthead overboard. It was believed that the creature had some foreknowledge of this calamity, waited to seize his chance and only then departed.

Animated with this superstitious dread, the Captain now sized up the monster shark alongside but then altered his mind. 'I won't spoil the edge of this on such a creature,' he said. 'He's all bone and gristle. Instead I'll strike his tail with the cutting-spade and see how he likes to cruise around without a stern piece.'

True to his word he took good aim and let fly with the spade. I fully expected to see the shark leave us in a hurry, but it did no such thing. It flicked its tail as though something had merely stung it and continued tearing at the whale — even though a large white gash was visible where the spade had cut the shark's sinews.

'Flames and flukes! I'll give him another,' the Captain thundered; and he did, but with the same result.

'That shark is like a rich man; he's got no feelings,' my companion sagely observed.

'Then I'll make him have some,' the indignant Captain retorted and continued spading the shark until, by repeated lucky hits, the tail was suddenly lopped off. The monster began to realise the want of such an appendage and, leaving the carcass, turned round and round and seemed unable to comprehend his misfortune. If he attempted to go in one direction, owing to the want of a rudder he steered an erratic course. His effort to regain the carcass was a lamentable failure, for instead of getting there he brought up among his jealous companions, who had been regarding his eccentric movements with some interest.

One, more bold than the rest, made a show of approaching and even had the audacity to put his square nose close to the severed tail. Others took courage from this example and one after another of the petty rascals made attempts to bite their big brother until the whole school was ranged against him and chased him from our sight.

'Well, isn't that just like our own world,' observed the Captain. 'As soon as you have lost all that is valuable, hundreds of erstwhile companions chase you out of favour!'

The tenacity with which a shark holds to life is astonishing and hardly to be credited by one who has not himself observed it. We caught a number for their skin which, cleansed and dried, is an excellent substitute for sandpaper and is much used in whaleships to smooth and polish the various things we make up out of bone and teeth. One shark that was hauled up on deck, even after it was cut open and its heart and other internal organs were removed, still continued to thrash with its tail. We were amazed to discover that the heart kept contracting for fully twenty minutes after it was taken out and pierced with a knife!

But to return to our whale and its bounty. The upper part of the whale's head was now slit open lengthwise, disclosing its case full of liquid spermaceti, clear as water. We bailed this out with buckets until, with nothing remaining of any value, we loosed the lashings and sent the great mass plunging downwards. It sank like a stone, eagerly followed by the sharks who were always hovering nearby.

As may be imagined, there was much oil running about the deck, for so saturated with it was every part of the creature that it had gushed out during the cutting-up process. None of it was allowed to run to waste, though, for the scupper-holes which drain the deck were all carefully plugged. And once again the whale contributed to extracting the last ounce of value from its life: the trailing edges of its flukes were cut into 'lippers' and we used them to scoop up oil from the deck, then added it to the try-pots.

Two harpooners attended these pots, replenishing them with minced blubber and bailing out the boiled oil into the great cooling tank on our starboard side. One officer superintended the mincing of the blubber, another exercised a general supervision over all. Because the ship was hove-to, there was no man at the wheel and no lookout needed to be posted: the fiery hell of our try-pots could be seen for many miles. And so we toiled watch and watch, six hours on and six hours off, the work never ceasing for an instant, day or night.

The whole scene had a wild and savage atmosphere. The murderous appearance of the blood-stained decks, the huge masses of once-living flesh, the ferocity in the looks of the men — all this was heightened by the hellish glare from the fires.

And then, after labouring amid the fires of hell, there was a singular bonus for some of us. After every sign of the operations was cleared away, the jaw was brought out and its teeth extracted with a small tackle. Scrimshaw is a favourite pastime on board every whaler — though I seldom saw much evidence of it being done, as hardly any leisure was allowed during daylight. Our carpenter, however, turned out to be a famous workman in whalebone and eventually completed half a dozen elegant walking-sticks.

A DEAD WHALE OR A STOVE BOAT

It must be admitted that our pursuit of whales was not always crowned with victory. Disappointment and frustration were often the only prize we took back to the whaleship. On one such excursion we were approaching whales in light weather and therefore using our paddles rather than noisy oars so as not to gally our foe. It seems that the whales' hearing is keen and that they are more alerted to danger by this organ than by their restricted eye. And so nothing could be heard but the bows of our boat cutting through the water and the whales ahead spouting. We all looked for fine striking and an easy target.

But just then one of the men struck the bottom of the boat with the edge of his paddle. Before the sound had ended, the whales disappeared like so many grindstones dropped into the water, leaving nothing to show but foam and bubbles here and there upon the surface.

Everything in the boats was quiet for a short time; then the air commenced to turn blue. Its colour changed so profanely that some said they could smell sulphur. The poor devil who had accidentally hit the bottom of the boat with his paddle, and so gallied the whales, wished himself behind an altar in some church so as to escape the oaths and curses showered upon him.

After we had lain to with the sheets of the sails slacked off for half an hour or so, the ship ran up the signal that the whales were in sight on her weather beam. Down came our sails and, with oars pulling, our boats were soon heading to windward in pursuit of our startled foe. As the ship was lying in our course we came within hail of her. The Captain sang out to us: 'The whales are about three miles off.' Although he didn't think we had much chance to catch them, as they were going quite quickly, he still urged us on.

We straightened out on our oars with not a smile on any face, as you may believe. About mid-morning the wind died out to almost a flat calm, the sun poured down its heat like a furnace and the perspiration streamed down my body and dropped from my heels. The others in our boat suffered alike — but our efforts were in vain. We caught no whales that day.

Indeed, it was to be nearly a week before we made fast to our next whale and so great was our determination to secure this prize that we all affirmed the whaler's most repeated oath: 'A dead whale or a stove boat.' But if we could have foreseen the day's unfolding, I doubt if our bravado and enthusiasm would have been quite so highly pitched.

With two irons home in our prey we strained to anticipate his next move, but when it came it was not as expected. After one or two glances around the horizon he calmly sank from view. We could see the men in the second mate's boat contesting every inch of line for fear that the whale should take it into his head to sound deeply. For a while it looked as though this expedient was to be successful, for the strain on the line seemed to cease. But while we were thinking that the whale was at least fifty fathoms beneath us, there was a sudden ripple and, to our consternation, the devil was right alongside and perusing us with his little eye as if we were now the prey.

'Stern all! Stern all!' yelled our mate. 'Stern all, I tell ye!'

Up out of the water rose the head with the high mouth massively gaping. Up, up it went until it towered above us like some high rock which needed but a touch to fall and utterly annihilate us. We forgot our oars. We forgot our boasting. Our whole horizon was filled with the overarching danger that threatened to overwhelm us. I looked to our mate for orders but he too seemed aghast at the immensity of the huge maw rising beside us. My resolve broke and I began to move in my seat and think how I might retreat from this disaster.

'The Devil may stay in such company,' declared one of the men in panic, 'but I won't.' And over he went.

I also wanted to go, but dreaded to leave without orders and all the while the mate kept mechanically repeating, 'Stern all — hard.'

'Here come flukes,' cried the boatsteerer and over he went as well. I needed no further inducement and struck out for my very life away from this impending horror. When I gained the surface I made for the third mate's boat which had pulled toward us to render some assistance. I dared not look behind me for fear of what I might see, but out of the flurry of noise a cruel splintering sound announced that the whale had closed his jaws upon my boat. It was impossible not to turn and witness his extreme rage.

I saw that he had divided it in the middle as easily as though cut through with a knife. Then the savage monster threw most of his body out of the water and came down upon the pieces to destroy even these last remnants. This was not a time for inaction, but with the whale now surging towards me I seemed marked for destruction.

The eddy which his huge body caused as it passed by surged me up against his flukes. I felt them touch my shoulder but as gently as a lady would tap her lover with a fan while in a sportive mood. Then the undercurrent drew me beneath the surface and I returned to a full consciousness of my perilous state. I resumed my desperate fight for life until I was gratefully dragged into the third mate's whaleboat.

Nowhere was our enemy to be seen but all around us the evidence of his attack was visible. Like a deserted battlefield strewn with the aftermath of war, the waters flung our pathetic wreckage back to us. I heard again our boastful cry, 'A dead whale or a stove boat', and cursed such foolish bravado.

AS IDLE AS A PAINTED SHIP . . .

For all these adventurous initiations into the life and experiences of a whaleman, it must be admitted that this first, voluntary whaling voyage of mine also saw more than its compensating share of uneventful days and placid sailing. For whole days and nights together we sometimes lay almost 'as idle as a painted ship upon a painted ocean', when the deep blue dome above matched the deep blue plain below and never a fleck of white appeared in sky or sea.

Except to a few who inhabit a kingdom of the mind, and others who can hardly be said to have any mind at all, the long monotony of unsuccessfully seeking for whales can be very wearying. The ceaseless motion of the vessel rocking at the centre of a circular world of blue, with a perfectly symmetrical dome enclosing all, can become almost unbearable. Were it not for the trivial round and common task of everyday ship duty, some of the crew must become idiotic or, in sheer rage at the want of interest in their lives, commit mutiny.

For the next two weeks we cruised in vain; not a whale could we find. The wind had been getting less and less and then one afternoon it gave out altogether. The ocean was like liquid glass. The heave it gave was so slight that the ship would hardly raise or lower the upper course of copper on her sides above or below the water. There was no bellying wind to fill the sails. Verily, it was silent enough for us to hear the small sound of bubbles coming from the nail holes in the copper sheathing as the desultory swell reached them now and then. Above the waterline more ominous bubbles could also be discerned: a lazy line of small black globules coursed down the sides of the ship as the pitch in her seams melted and was drawn out. With these sounds our whaleship seemed to be like some mighty breathing beast itself, becalmed, alone and forgotten.

Below, in the fo'c's'le, the heat was almost intolerable. We lay in our berths, trying to sleep, with perspiration streaming from our bodies. Those on deck sought out the little shade afforded by the masts and sails, but those at the mastheads had to endure the sun's cruel affliction. The mirror-like ocean reflected these burning rays and made it very hard on the eyes. When the sun went below the line of the horizon, a sigh of relief was heard from the knot of men around the windlass.

Our sails hung dejectedly from the spars, as downcast as our spirits and our stamina. We could only detect some small motion of the ship from the compass. Although it seemed we never moved, we could see from this instrument that sometimes during the night our ship would turn completely around. But so imperceptible was its rotation that we could never feel the welcome breath of a turning breeze.

It had been many days since we took our last whale and this lack of activity was the greatest trial that could attend us. There is no mode of life, it appears to me, requiring so great a variety of talent as whaling. Every one of us must be a 'jack of all trades'. He must have mechanical talent sufficient for making all repairs upon the sails, rigging, iron and woodwork of the ship, and as he is absent from port for months together he must have a talent for shoemaking and tailoring. Coats and pantaloons which others would pronounce to be 'unseaworthy', they work upon and cover over with patches of various colours until not a vestige of the original is left. The performance of the men with their needles is by no means contemptible. But now even the demands and distractions of these household chores had been exhausted.

The dullness and tedium of life on board ship at such quiet times can be almost unendurable. The harpooners incessantly checked their gear, sometimes spending their entire morning sharpening harpoon and lance points that only the previous day had been made as sharp as any whetstone could hone them, or re-coiling lines that already lay perfectly within their tubs. For the rest of us, the daily routine — make sail, man mastheads, scrub decks; breakfast, dinner, supper; shorten sail, boat's crew watch, 'turn in' — could not provide a single line for any journal. The men had become morose and quarrelsome, numerous scores were run up and appointments settled to fight them out in the first port we made. The violin failed to move, the song to enliven, or the yarn to interest us.

According to custom, and as a diversion, a red flannel shirt had been offered as a prize to him who raised the first whale and then a pair of duck trousers had been added. Pounds of tobacco were offered by the mate but the days passed monotonously. A gold doubloon was nailed to the mainmast as another reward for good eyes. With this incitement not a whitecap could show, a porpoise jump or a finback spout but that the alarm would be given. Yet all was in vain.

So desperate for diversion were the crew that one even found idle amusement in baiting an unwary albatross. With great confidence I was assured that these birds were so oily that it would make me waterproof to eat them! I was not tempted by this offer and took more pleasure in admiring the magnificent spread of their wings. I have seen their long hollow bones provide ornamental needle-cases for sweethearts back home, although it seems a trivial use for such an imposing prize. This particular catch, however, was to be preserved from death and converted into a messenger.

Using a floating hook with some pork fat, my companion eventually snared one grand old bird whose stretch of wing must have been twelve feet, or near enough. Writing the words: '— of Hobart-town, Captain Meredith; all well; latitude — S longitude — W', he secured his message in oiled silk. This was then varnished waterproof and secured by a cord around the hapless bird's neck. A red ribbon was attached to attract attention and, all being prepared, he launched his messenger to set a true course for sweethearts and wives with a loud hurrah.

UNDER THE LEE OF THE DEVIL

One memorable exception that startled us out of any feeling of monotony was a violent tropical thunderstorm we encountered. At this time we were in the 'horse latitudes' and, much to our relief, the rain came down in deluges, permitting us to wash well and often. I suppose the rains of the tropics have been often enough described to need no meagre attempt of mine to convey an idea of them, yet I still wished I could make home-keeping friends understand how far short what they call a 'tropical shower' is from the genuine article. The nearest I can get to it is the idea of an ocean suspended overhead and then cascading its wrath and majesty down upon us. Nothing is visible or audible but the glare and roar of falling water, and a ship's deck, despite the many outlets, can be quite submerged.

For the past day or two the weather had changed and the sky had lost its clear bright blue, which had been replaced with a reddish mist, somewhat like a fog but without dampness, through which the sun shone blood red. It was as if it were shining though air filled with smoke and the rigging and sails cast shifting shadows on our deck. It so looked as if we were sailing into the very gates of Hell that the continual preternatural glare made us distinctly uneasy. Our mood was not lightened when, now and then, a faint moaning sound could be heard in the rigging.

In the early evening the Captain came on deck. The sky had shut down thick and dark with not a star to be seen and the very air seemed oppressive and not free to breathe. He walked the deck a few times and then gave the second mate orders to close-reef the main-topsail, furl the fore and mizzen topsails and see that everything loose around the decks was firmly lashed. 'If I miss not the look of this weather,' he confided, 'we shall catch a hurricane before morning. Let me know if any change takes place.' Then he went below.

As the night wore on, the swell grew more heavy and the air closer. On going below, the steerage seemed so hot and close after turning in that it was hard work to get to sleep. At three o'clock, when my watch was called, the ship had such an uneasy motion to her that I stumbled about dressing. When I reached the deck the darkness was intense. The ship seemed in a world of her own. Heavy black clouds hung above us so low that the masts seemed to penetrate them. Nothing could be seen half a dozen ship's lengths away from us in any direction, except the dark waves whose tops now and then would break with a spiteful snap, momentarily showing a burning phosphoresence.

The ship was rolling fearfully, both to windward and to leeward. Now and then she would pitch her head down, almost burying her jib boom under water, and raise it again with water streaming from her bows in torrents. Then her stern would settle with a shuddering force before she was almost far enough below the surface of the water for it to come over the taffrail. It seemed that something must give way before the terrible force of the sea, which at times made the ship tremble throughout. When her masts and yards performed circles and abrupt angles in the air, she seemed like a distressed person tossing her arms about for help. And if by inattention or carelessness the man at the wheel should lose hold of it, most likely it would fly around and break his arms before he could get out of the way.

All of us were frighteningly aware that something out of the common was about to take place. Now and then the Captain's face could be seen in the light from the binnacle when he glared at the compass, and as a fitful dawn broke he told the second mate to have the men stand by the main braces. 'Never mind coiling the ropes,' he said. 'We shall soon have on us a hurricane that will try our souls and our good ship likewise. I have no doubt it is coming from where that light shudders ahead.'

He had hardly finished speaking when the rag of the main-topsail slapped hard aback against the mast. This sudden wind screeched through the rigging and then was just as suddenly gone. So quiet was everything after it ceased that the order from the Captain to lay square the mainyard startled us all. Hardly had we completed our task when, with a roar like a mad bull, the wind struck us. The ship's stern commenced to rise and her bows to settle down and down and it seemed that she might dive into the depths of the ocean like a whale turning flukes. Hanging this way for some little time, as if undecided whether to go to Heaven or the other, she gracefully settled her stern and sent half a deckload of water rushing aft.

The whole celestial machinery of rain-making was observed in full working order. Five or six mighty waterspouts in various stages of development were seen within easy distance of us; once, indeed, we watched the birth, growth and death of one less than a mile away. First a great assemblage of black cloud began to belly downward until the centre of it tapered into a stem and the whole mass looked like a vast irregularly moulded funnel. Lower and lower it reached, as if feeling for a place to anchor, until the sea beneath was agitated sympathetically, rising at last into a sort of pointed mound to meet the descending column.

So close were we, it was possible to see that both descending and rising parts were whirling violently in obedience to some invisible force. And when they joined each other, the upward rush of the water was very plainly to be seen. The cloud overhead grew blacker and bigger until the bruised sky was fearful to behold. Then the spout gradually grew thinner until it became a mere thread, and we could not determine when the connection between sea and sky finally ceased altogether.

Our ship ran fine but the wind came so hard that at times it almost blew us clear off the deck. It blew so strong that when the men were about ten feet up it pinned them tight against the ratlines and you could not have made them hear without a trumpet. The wind was howling with such force that the sea would not rise. Instead, it was rolling over with a white foam; and the foam, as it dashed against the weather side of the ship, would send a spray over us like fine drifting snow. The hurricane increased in fury until the ship was pressed bodily down into the water and held there.

Our main-topsail sheet parted and slatted itself into ribbands with reports like a cannon that could even be heard above the howling gale. It left nothing but a row of matted threads of canvas on the yardarm. And then, most welcome relief: the wind began to drop and our ship to stand more upright. Quite suddenly, it seemed, the sea was calmed and a thick mist rolled in. If such an occurrence had happened at any other time but now, it would have caused great wonder, but all of us were exhausted and our worst expectations keyed almost to breaking point. Those among the sailors most easily gallied and prone to superstitious dread observed an omen in that hollow ominous flap of sails against our masts and interpreted the unsettling vibrations in the cordage as presaging something dire.

In the calm respite our ship became covered with birds which had been blown off course — bright-coloured waifs so wearied by the storm and so desperate for a place to rest that they forgot their natural fear of man. However, this was our chance to clear the forlorn sails from the bows and stays and replace them. Our exhausted passengers reluctantly flew off with many sad and complaining cries of distress.

The ship had such a dreary look, but the change from the incessant noise of the tempest to the quietness now was almost too wonderful for belief. No man on the ship had seen the like before. The Captain could not fathom what it meant to have the sea so suddenly grow smooth. The wind almost going down completely was nothing, that was often the case; but the other part beat him, he said.

The mate spoke up: 'I don't know any way of accounting for it, except we are under the lee of some land.'

'Under the lee of the Devil!' the Captain exclaimed. 'Where is this land? We could almost have been blown right over any islands hereabouts.'

But then the mist began to clear. The sun was seen shining through and even the horizon could be discerned. Someone remarked that it seemed strange the mist was clearing faster ahead and astern than on each side of us, where it seemed to hang, looking more like murky land than fog banks. At that the Captain's voice startled us all with a loud bellow. 'Flames and flukes,' he roared, 'here is land on both sides of us. How in Hell is this?'

Everybody stopped work and, looking off both larboard and starboard beams, could see land at no great distance away! By some good fortune our ship had got between two islands and into safety under their lee side. I have often thought since of the horror that would have come over us during the storm on seeing those rocky shores only just before we struck. No one on board would have thought but that we had plenty of sea room. No lookout could have been kept and we would only have known our fate as we met it head on.

Perhaps it is best to go that way, but it seems to me that a horror sprung on a man like that would have been a lifetime of agony contained in a minute. I, for one, had rather see a little longer ahead.

At dawn the next morning we found that we were not the only ones who had found a haven in these sheltered waters. Anchored as we were close to shore and hardly expecting a prize to appear, imagine our surprise when we distinctly heard a long expiration and beheld a lone white form shimmering against the dark mass of cliff not a quarter of a mile away. Taking one boat out and silently resorting to our paddles rather than any noisy oars, we drew on through the dim morning light.

Here we found a cow and her calf.

Close nestled to the side of the whale was her young suckling of not more, certainly, than five days old. One long winglike fin embraced its small body, holding it close to the massive breast of the tender mother, whose only care seemed to be to protect her young, utterly regardless of her own danger. The cow could easily have fled but would not leave her calf. She tried every expedient to quicken its progress, first bearing it on her flukes, then taking the little thing between her fin and her body. When she found this useless, she placed it before her to push it along with her nose.

But all was in vain. The mate pulled up and, by way of making sure of the mother who had just gone down, put an iron into the calf. Our harpooner rose, darted once, twice, and then gave a cry of triumph that rang re-echoing all around in a thousand eerie vibrations, startling a host of noisy birds aloft. As soon as the mother came up she was fastened too and in a few minutes was spouting thick blood.

But this was not like any other catch we had made. For all the notice taken by this whale of our barbs, she might never have been touched. And then as the light grew and we pulled in closer we found reason enough for this uncomplaining suffering. She seemed solicitous only for her calf. She would get between it and the boats to keep it out of further harm's way. She would take it down with her knowing that on the bottom was the safest place. But here the little one could not obey her. It was forced to surface to breathe at least once every two minutes. By this means, even if we had not been able to tell by the strain on our lines, we knew at all times just where the mother would surface.

Truly, if sentiment were ever permitted to interfere with our operations it might well have done so now; for while the calf continually sought to raise comfort from its mother's enfolding fin, making all sorts of puny struggles in the attempt, the mother scarcely moved from her position. And all this while she streamed with blood from her wounds.

When we were scarce half a mile from the ship, a sea breeze came in and, as she felt it, new life seemed to enter the mother. She again tried to push her young one ahead. Finding this useless and that we took occasion meantime to ply her with lances, she suddenly turned upon the boats.

When one would approach her, she would whirl about and dash at it with her flukes. By continually running around in a small circle she got our various lines so fouled that we began to fear we should have to cut from her and fasten afresh. Once, indeed, as a deep-searching thrust entered her very vitals, she raised her massy flukes high in the air with an apparently involuntary movement of agony; but even in that dire throe she remembered the possible danger to her young one and laid the tremendous weapon softly down upon the water as if it were no more than a fan of feathers. And then, in the most perfect quiet, with scarcely a writhe nor any sign of a flurry, she died, holding her calf to her side until her last animating spark had fled.

Case-hardened hunters as we were, we all felt deeply ashamed of our deed. With guilty eyes we looked at each other and then quickly away, little caring to see the useless pity only now displayed there. It needed but one merciful thrust from a lance to end the calf's misery. Truly, no slaughter of a lamb ever looked more like murder. Nor, when the vast bulk and strength of the whale was considered, could a mightier example have been given to us of the force and quality of maternal love.

It seemed strange that we had been saved from the very jaws of death ourselves only to regain so quickly our killing advantage over these creatures. We could not be proud of this deed.

Part Three

DRESSED FOR THE PART

Back in Sydney-town I was again returned to the welcome embrace of family and to the daily comforts of a more settled existence. And it was then, after my next voyage and as I approached my twenty-third year, that I first began to steer a new course, one navigated by the beacon of another's heart. In short, I began to value my days ashore even more than my days at sea, for it was now that I began courting the young woman who would guide me into my life's most blessed mooring. Susannah's arrival in my life was as unexpected as it was arresting.

On my last voyage out of Sydney-town in 1826 there was the usual complement of greenhands. Among them was one young man who displayed a refined bearing and quiet deportment which had seldom been seen before in the fo'c's'le of a whaler. It was immediately obvious that, although he had signed on in the loose and democratic attire of a sailor, he was clearly a gentleman. Such rare birds of passage were not entirely unknown on whaleships. Disgraced sons, foolhardy adventurers and the like have their own reasons for fleeing a family's fond embrace and seeking the solace of the sea.

Once this greenhand had recovered from the usual preliminary bout of seasickness, we could find no fault in his willingness to learn the ropes and observed the alacrity with which he applied himself to his duties. When reefing topsails he never seemed to heed how hard the wind was blowing or the sail flying over the yard. In the whaleboat he was always seen to pull his oar with energy, although, it must be admitted, his slight build hardly gave his strokes much effect. Still, by showing no more fear of the whale than the other greenhands in all things, he won the respect of his fellow shipmates and the goodwill of myself and the other mates.

We had been out about three months with the weather very hot and our young sailor had been sick for the past few days. One night, during the middle watch with everything calm and shipshape, one of the crew was heard bounding up the fo'c's'le steps. He rushed aft to me and, as the mate on duty, I must confess that I was alarmed by his abrupt appearance. I expected to hear some disquieting news of trouble with the ship. Imagine my astonishment when what I heard, and indeed what the whole deck overheard, was: 'That there fellow, that young fellow sick below, is no fellow. He's a woman!'

Since only the day before the Captain had come down heavily on one of the crew for brewing a wicked concoction of ale, I rightly assumed that here was another who had enjoyed a secret drink and was now paying the muddled consequence. 'What in Hell do you mean tearing up here in the dead of night with your foolery?' I said. And matching actions to words I startled the sailor still more by offering him my raised fist. The man backed away somewhat, but only to repeat his words which now had the attention of all on deck. 'It's true, sir. He's a woman. Come forward and you will see for yourself.'

Passing forward through the men, who were understandably eager to follow if there were any truth in this fabulous claim, I went below. Here I was directed towards a berth that had its curtain partly drawn aside. The only sound down here came from the fitful sleeping of the men in their overheated quarters. The light from my lamp was somewhat sickly, but even its meagre glare was sufficient for me to discern the truth in the man's claim. There before my astonished gaze, sleeping feverishly and thus revealed in a dishevelled female state, lay irrefutable evidence that this was indeed a woman!

It hardly seemed credible that a crew member should have been surrounded for months past by over a dozen men in close proximity and still have retained such an inviolate disguise. I removed my lamp, closed the curtain with a considerate gesture and only then called the woman by her masculine name.

It was enough to waken her. There followed a muted exchange through the curtain in which I told her that her secret had been revealed and that she must dress and come aft with me. A stifled gasp was all that issued from the berth as I discreetly withdrew and waited in some confusion. Then we proceeded to the forward cabin where our Captain was called from his stateroom. When he could finally be brought to believe this novel situation, he offered the distraught woman spare quarters in the after cabin. In response to this kindness she volunteered her story.

It seemed familiar enough. She had formed an attachment to a young sailor from a whaler and had mistakenly believed that he had signed on to our vessel. Being of a determined cast of mind and not choosing to stay at home, she devised a daring if foolish plan. Being quite tall for a woman and built slim, she donned men's clothes and easily passed herself off as a greenhand with a keen desire to go whaling.

Throughout the telling of her story, I must admit, I kept recollecting events while she had been in disguise which caused remarks at the time. She had been seen to blush so often at rough speech made in her presence that it had finally become accepted that this greenhand had led a very sheltered life. It is true that she had been reserved and even in the close confines of the ship had managed to keep to her own company much of the time. One of the mates remarked, later, that he remembered an event which now gained new significance.

He had observed her washing clothes and, noting the shipshape manner in which it was being done, had paid her a passing compliment: 'By the way you wash those clothes it would seem you were used to taking care of your things.' The mate now recalled that his slight praise had evoked a quite unexpected reaction. She had flushed bright red, hung her head even lower over the wash and made no reply. And when she resumed washing he observed that she only did so in as awkward a manner as the most ignorant greenhand on board. After our discovery and when we cleared out the bunk which she had occupied forward, we found a piece of topgallant duck that had been cut to fit her form, lacing up on each side under the arm. By drawing this tight she would flatten her bust and keep her sex a secret.

When our young renegade somewhat shyly made her first appearance on deck in her true guise, you would have thought a fabled mermaid had come into our midst, so great was the interest. At first we were struck dumb with astonishment when our former crewmate was brought forward by the Captain and introduced as Susannah Andrews. Over the ensuing days she fashioned more feminine attire, until it was only her sunburnt hands and short hair which gave any clue to her former disguise. But even so, it took us some time to adjust to this enticing young Venus addressing us all by first name.

To say that all of us men aboard were much distracted thereafter by the presence of this spirited young woman would hardly reveal the intense rivalry that operated to catch her eye or perform some small service for her. We were all determined to show that although her favoured sailor was not among our crew, any one of us would gladly deputise in his place.

When it came time to say farewell, we all came up on deck. Here we assembled and, one by one, shook Susannah by the hand with earnest protestations of good fortune. Throughout this ordeal she showed her true mettle by looking each of us in the eye and bidding us a caring goodbye. I cannot doubt that each and every one of us wished himself her champion. If her former lover had come on board and been revealed to us just then, I truly believe that his life might have been forfeit!

Even today, thirty years later, I still hear versions of this story recounted and must confess that I do not admit to the appreciative audience that I was one of those young whalers duped by a female greenhand. And as much as she may smile indulgently at my pride in her resourcefulness, once Susannah became my wife she never again evinced a desire to set foot on board a whaling ship.

A NEW EDEN

At this time, when the rage for emigration had become so prevalent and when the most alluring prospects were daily held out by enthusiasts inviting us to wherever a new sort of Eden had sprung up, my eyes were directed to Van Diemen's Land. I was delighted with all that I witnessed there and I saw singular advantages for our whaling industry in this adventitiously sited island. There was also the alluring prospect of establishing a home for myself and Susannah, to whom I had now been affianced for some months. Caught up in the heady promise of life in this new Eden, I lost no time in writing to Susannah.

William Chamberlain to Susannah Andrews, Sydney.

Hobart-town
May 1827

My dearest Susannah,
Although I write to you separated by so many miles, still I feel a closeness to you and a great joy in knowing that soon our voyaging

will be as one. Your dear image is as foremost in my mind as your well-being is in the vanguard of my emotions. Do not think me flippant when I tell you that I am eager to explore the harbours, the islands and the gentle sailing along the coast of the land of matrimony. Already, you see, I overlay the real landscape here with plans and fond dreams for our future together.

So, let me tell you something of this picturesque location. The Derwent is a noble river, very wide and navigable for one hundred miles. As for the Town's harbour, one views its commodious expanse with admiration. On the other hand, it is true that there is but one major road across this island. It reaches some miles out of Hobart-town and Launceston but it has not been so far extended as to join in the centre of the island. Yet for that reason it makes coastal trading and shipping a lucrative proposition. The island has a fertile and luxuriant soil; a healthy and [not withstanding its inconstancy] a salubrious climate. And lest you think this distant colony uncivilised, I must say that its wilderness is being subdued by husbandry.

With all these recommendations, however, I will admit that I am glad I am a whaler and not a farmer. Some of our crew, on contrasting their present rough and boisterous calling with this more peaceful one, are rather disposed to think that farmers have the best of it. Several therefore expressed a willingness to exchange conditions with them. But the latter's way of life is decidedly not to my liking and I would sooner plough the uncharted leagues of ocean any day, than the most verdant pasture here.

The farmer has convicts to help him, but it is sad to say that his expenses, instead of being diminished, are actually increased by having convict labour. The men saunter through their time and work with reluctance. Most of them have been bred in towns and hence one finds London pickpockets as shepherds, pastrycooks as ploughmen and chimney sweeps as bullock drivers. Insolence must be forgiven, intemperance overlooked, distinction shown, presents made and holidays granted — and all to reconcile, frequently, a sullen fellow to his daily, easy and overpaid duties.

As for the convict women, I have always wondered if they are the proper persons to send as servants into private families. Can a master or mistress be constantly watching over the actions and conversations of his or her servants, in order to save the children's minds from contamination?

More optimistically, however, the chances of the convicts' reformation are rendered more likely by their residence in this small corner of the world. In a town which numbers just over 3000 inhabitants there are wages and employment enough for anybody willing to work. Idleness and want, therefore, which are the main incitements to vice, may cease to operate. The convicts have no longer the same opportunity of pursuing those evil practices into which they have been led by bad company in large towns with no prospects. When want is out of the question, the choice between virtue and vice should be comparatively easy, and is certainly much facilitated by the parental care of Government.

The industrious and well-disposed are encouraged by being allowed to go out as servants to respectable settlers; and if, after a certain period of good conduct, they have maintained such a character in the neighbourhood where they have resided as to warrant the Governor in making them free, he not only confers that favour on them but allows them a grant of land.

But enough, lest I turn your opinion of this small paradise away from my intended goal. To look beyond to the broader horizon — this is just such a place where together we might be most comfortably settled. Dwellings having been erected long before the streets were made, and the town being upon an irregular surface, some of the buildings in consequence now occupy very awkward situations. On one side of a street they are often elevated much above the level; while, on the other, they are sunk considerably beneath it. As every proprietor has been guided by his own taste in the structure of his house, few are built alike or upon the same plan, and as he is not restrained by the Government to a settled line, they are planted too often in zig-zag positions.

It occurs to me that if we settle here, I could join this eccentric company when we build our own house. I have long harboured a desire to replicate the viewing platform which my father built onto our home in Sydney-town and no one would look askance here at such a folly. But before my sails become too filled with enthusiasm, I would welcome your opinion on a proposed southern migration.

This is a community which is rapidly developing. For two years it has been a Colony separate from New South Wales and I believe that nothing will advance it more than the Hobart whaling industry. Many of the Colony's bays and harbours abound with right whale in the month of November. At this time they are on the point of calving and take shelter here. They remain with their young between two and three months and the inhabitants of Hobart-town are frequently able to witness the whole method of killing them from the vantage point of their very own windows.

As a master mariner I foresee a future and a living that offer much enticement. I hope that together we might settle upon Hobart-town as a congenial place to raise a family. I am certain that it offers me the opportunity to chart a clear course into the whaling industry. The sea-going fraternity has welcomed me into this bustling port and I could easily conceive myself rising to master of my own ship within a decade.

I am told that the state of Society has been gradually ameliorating by the accession of a great number of respectable settlers, some of whom have brought well-bred horses, cattle and Spanish sheep with them. The coursing of the kangaroo and emu furnishes sport and exercise equal to the stag-hunting of England. Kangaroo meat is regarded here as equal to venison and the tail makes excellent soup, quite superior to ox-tail.

The flesh of the emu is also considered wholesome and the birds are easily tamed. I have seen one householder who has a pair that perform the office usually allotted to a house-dog. If a stranger intrudes himself, the emu interposes, and if the fellow does not quickly retire, the emu will not fail to throw him down by a push with his foot against the head or breast.

In conclusion, my dear Susannah, I believe that there is safety, opportunity and congenial society here in Hobart-town . . .

THE *FREDERICK* SAILS AGAIN

And so it was that a new chapter in my life's voyaging began. Never during the whole period of my adult life did time pass so happily as when I navigated the shores of matrimony. For years no cloud presented its dark shadow, no sorrow gave a shade of gloom, no violent storm wrecked our calm sailing. But then a presentiment came early in 1842, when the dreadful spectre of the *Frederick* was revived in my memory.

Only the year before I had returned from a voyage eager to recount a memorable meeting at sea with a Yankee whaler and his happy evidence of domestic comforts afloat. These Yankee whalers are a formidable crew of men who have made a significant impression on me over the years for two very different reasons. The first is that many of them are Quakers hailing from Nantucket on America's east coast. My surprise at these meetings came from witnessing such godly men, who eschew any form of violence to man, proceed to do battle with the whale in a display of single-minded ferocity and determination. One of them explained their philosophy as a rational mixture of enterprise and religion. 'We would never forget the Lord's day. But give up whaling on a

Sunday and miss a fair chance? That would be to reject Heaven's bounty!'

Their other surprising characteristic is their fond practice of occasionally bringing their wives to sea with them. Of course, their voyages are always of much longer duration than ours; sometimes, indeed, they may be away from home and family for as long as four years or more. Then it might not be unusual for them to arrive home to find a new child not just born but walking and talking. And this infant newcomer will have no idea whence comes this sudden large stranger who takes away much of the attention of his previously doting mother. Susannah and I now had four children. The eldest, named after her mother, was almost twelve and the youngest almost four. I could not imagine myself at sea and gone from them for longer than the usual five months we spend during the whaling season.

I admit that I was astonished to see the way some of these Yankee whaling wives go to sea. That they turn a rough captain's stateroom into a domestic haven is not the only remarkable feature of this practice. It is also apparent, and not surprising, that a female presence on board significantly ameliorates the harsh sentiments and stern dispositions which men can assume when deprived of all female company and refinements. But for all this I still do not think that the whaleship is a fit place for women. It is a savage stage upon which a war is waged against the elements and against the Devil that is in the sea. And yet, on my return home, I could not help telling Susannah of my meeting at sea with the Yankee captain and his wife.

These remote meetings or 'gams' are a pastime peculiar to our whaling profession. Merchant ships will pass each other in mid-ocean without a sign of acknowledgement, but whaleships, when they meet, are friendlier. They will heave to after the day's cruising is over and there is no longer any chance of raising whales, and the captain's watch of one ship will entertain the captain's watch of the other.

I could tell, even as I described the domestic comforts which these Yankee captains could still enjoy away from home, that my dear wife was imagining herself in a similar position. Lest she felt she was being compared unfavourably, I hastened to amend my praise of this Yankee couple. For in truth, my wife had always been watchful of my comfort on these voyages and any departure was a season of anxiety and preparation. Mattress and bed-clothing, trousers and jackets, stockings and shirts, pea-jackets and storm-coats would all be carefully overhauled and any rents in the garments made whole. New supplies of clothing would be added to suit all weathers and special keepsakes stowed away in chests and clothes-bags that would later reveal the tender consideration of a wife for her sailor husband.

Susannah often claimed that she had two husbands — the one who was hers at home and the other who belonged to the sea. There was, of course, truth as well as good humour in her claim. Yet if a man is going to have a mistress, the Pacific would be the most tempting and seductive that I could imagine. But is the embrace of the sea greater than that of a wife?

Good humour and contentment like this filled our days. But then, in the midst of years of calm sailing, the family proved to be aboard a more fragile vessel than any of us had imagined. Suddenly our contented life was swamped and our happiness swept overboard. We suffered a terrible loss and during our grieving I could not, for any prize at sea, leave my family to go whaling as usual.

Our eldest child, our Susannah, named after her mother, was suddenly taken sick and died in 1842. We lost not just a daughter but a mother's fondest companion; one who was always so caring of her three younger brothers was suddenly no more. The shock was so great that I resolved to forgo part of my next whaling voyage entirely and spend some time ashore.

It was then, in that vulnerable state of grief and concern, that the name of the *Frederick* was revived in our daily lives.

Our first intimation of disconcerting news arrived in the pages of one of our Hobart newspapers. It reported that a Government brig, built by convict labour about 200 miles away on the west coast at Macquarie Harbour, had been piratically seized by the same felons in mid-January. And the ill-fated name of this ship? None other than the *Frederick*.

In all my years of whaling I had never once encountered a ship with this same abhorrent name. For a new vessel so near at hand to be launched on its maiden voyage so nefariously and to bear the name *Frederick* was, I feared, a dreadful omen to myself and to my family. Macquarie Harbour has a notoriously forbidding entrance, aptly named Hell's Gates. But while we imagined that this threatening bulwark could contain our desperate convict population, it had utterly failed. Once more a ship with the evil name *Frederick* had been unleashed on a hellish voyage. I feared that a devil was stalking my days again.

It appeared from the newspaper report that the prisoners at Macquarie Harbour had taken advantage of some of their military guards being absent on a fishing expedition. That the convicts had overpowered the remaining soldiers and taken forcible possession of the very ship they had just completed building without resorting to murder was the only point in their favour. Previous to this, the soldiers had had a good opinion of their convict charges and no apprehension whatsoever of their attempting any such act of piracy. The whole establishment at the Harbour was, in fact, on the very eve of disbanding and the convicts had chosen their time with cunning and enterprise.

All Government and private property that was considered of any use and value had already been shipped on board the *Frederick* so that its abductors could escape with a well-stocked hold. That such an act of criminal effrontery could be perpetrated almost on our very doorstep was an unsettling surprise to all of us. Where the felons were headed and what desperate plans they were considering became the subject of our daily conversation and apprehension.

To my mind, the resurrection of the name *Frederick* was sufficient to unsettle twenty-five years of secure life and more than a decade of domestic tranquillity. Indeed, the convict establishment at Macquarie Harbour had always seemed destined to arouse the most dire feelings of apprehension and terror in our community.

Seven years before Susannah and I had settled in Hobart-town, at the very beginning of 1822, the Harbour was selected as a place of ultra-banishment for the most violent and debased convicts. It was my colleague, the renowned whaling captain James Kelly, who had charted and recommended the site. His opinion on the impregnable natural fortress it presented greatly encouraged the authorities in their choice of location. And the selection of one small island even further isolated within the great moat of the Harbour should have guaranteed that it was fortress enough. But even before its first year was over, the place became known for fostering the basest of deeds.

Late in 1822, Alexander Pearce and seven other convicts escaped. After a nine-week trek, however, it was only Pearce who was returned to captivity. This sole survivor had reverted to his utterly bestial nature and henceforth Hell's Gates became synonymous with cannibalism and depravity. This then was the place's opening scenario; now it was closing with an equally notorious finale.

SCRIMSHAW

It was during this sad time following our daughter's death that a welcome distraction was afforded by a novel visit to the theatre — the Royal Victoria Theatre in Hobart-town. The occasion promised to be a unique one in our Colonial annals for it was a specially written drama commemorating a topical and exciting event in our own history. Before this I had not heard of any occasion in our small part of the world so honoured by a stage presentation.

Only the year before, the British South Polar Expedition, led by Sir James Ross with Commander Francis Crozier as second-in-command, had spent three months in Hobart-town before their brave foray into the southern unknown. The Expedition had been the constant subject of news reports, not least because its Captain enjoyed a close friendship with our Governor, Sir John Franklin. He was himself a polar hero and so keenly supportive of the Expedition's scientific pursuit that Government House was 'en fête' for much of the Expedition's visit.

Our *Hobart Town Advertiser* had been filled with the exciting plans and possibilities of the Expedition during its initial three-month visit. Its leaders had resided at Government House during this time and press reports of fulsome vice-regal hospitality and of reciprocal gestures aboard the Expedition's ships put Hobart society into a state of considerable and sustained excitement. It was reported that Sir John and Lady Franklin dined on board the *Erebus* with great ceremony; their arrival being marked by a salute from the ship's guns and rousing cheers from its manned yards.

We all felt that the visitors conferred great distinction upon our remote part of the Empire and everyone vied to secure an invitation to one or another levee or evening supper with them. Then, in late April 1840, the Expedition returned to us after successfully circumnavigating the icy continent. Now an even greater level of excitement was generated throughout society, which was to culminate in a grand nautical drama extolling the exploits of the Expedition.

After the performance we attended a supper hosted by Mr Cameron, the theatre's lessee and author of the evening's piece. Once there we were introduced to Captain Ross. 'Captain Chamberlain,' he remarked, 'I would be interested to hear what a local seafaring man thought of tonight's presentation.'

'I was much impressed, sir, with the depiction of the polar scenery, and even though I am a captain myself, the staging of your shipboard adventures certainly put me, for one, rather convincingly at sea. I had not expected such sound and lighting effects in this impressive production.'

My praise was gratefully acknowledged by Mr Cameron, who assured me that the scenery was as authentic as possible and was based on sketches made by officers of the Expedition.

Our host then prevailed upon Susannah for her opinion. She smiled at him. 'Mr Cameron, you must be highly gratified with the attendance at tonight's performance. Will it encourage you to plan other productions with such rousing Colonial interest?'

'The problem,' he replied, 'is to find local themes sufficiently novel and spirited enough to appeal to our audience. Certainly not something which simply rehearses once more our convict parentage — but a theme more elevating and patriotic.'

Susannah nodded in agreement. 'And might I add to that requirement, something with an acknowledged appeal to the feminine side of the audience. You will surely have noted how many women from all classes of society enjoyed tonight's performance?'

'And not just the women in the audience,' I hastened to add, for the night's presentation had certainly awakened in me a proud and patriotic reaction. 'I must also compliment you on the introduction of Sir John Franklin into the piece, an introduction which was managed without violating any decorous propriety. The slightest error in this particular would have destroyed the piece, let its merits otherwise have been ever so great.'

This well-deserved praise won Mr Cameron's hearty thanks before he continued to expound on the problem of finding inspiration in local themes. Some short time later I observed Susannah engaged in an animated conversation with a woman of quite striking presentation across the room. I gave thanks that this change of scenery seemed to be lifting Susannah's spirits and turning her thoughts away from our grief. Later I was to realise that my dear partner's equal concern for my own solace was prompting her into a surprising venture.

Over the course of the next year our family renewed its caring bonds, especially once we found that Susannah was again expecting. During the last months of her confinement she was much comforted by visits from the new friend she had made in Hobart-town after our excursion to the theatre. This companion was Elizabeth Cameron, wife to the Royal Victoria Theatre's lessee, Samson Cameron, and the two women spent many hours together and maintained a lively correspondence.

This was to be our seventh baby and I determined not to be a stranger to the new child. In the twelve years since our first was born, I had not once been at home and on hand to witness Susannah's pain or share in her triumphant first days of mothering.

When I was a younger man the matter appeared to me in a different light, but as I grew older my desire to stay home with my family increased and it seemed like tearing my heartstrings to depart on a cruise with the probability of being gone up to five months. During all my voyaging I would receive no knowledge of events at home. Indeed, by my calculation I have spent much more than half of the last decade at sea.

Susannah has always conceded that I enjoy myself more in the few months I remain on shore after my return than I would had I renounced the seafaring life altogether. And perhaps these long absences give a zest to the enjoyment of home pleasures which nothing else would impart to them. But I also realised that while I had never been present at a birth in the family, I had, most sorrowfully, now been with them for the death of a grown child.

In my absences Susannah had also birthed two daughters only to lose them within days; while I, away for months at sea, knew nothing of her own hard voyaging or its cruel endings. And so, as I waited out this confinement, I was for the first time aware of the many necessary preparations which Susannah and the midwife were making. The calm and steady course they set throughout these last weeks before the birth gave me a new appreciation of home and its comforts.

I soon realised that although the 'captain' of our household might be indisposed for some days or even weeks after the event, past experience and careful organisation would ensure that all remained shipshape. Ironically, it was I who felt somewhat left out of the women's bustling and it was then that I hit upon an engrossing task to fill the waiting hours with a quiet labour of my own.

During my most recent voyage, the lower jaw of the whale had been hoisted on deck as was usual and lashed to the bulwarks near the bow. Once every sign of the cutting-in had been cleared away, the jaw was brought out and the teeth extracted with a small tackle. These formidable ivory ramparts are set solidly into the white gum, which has to be cut away all around them before they can be retrieved. Together with the hard white whalebone they were cleaned and then headed up in a barrel of brine. Now they were ready for eager sailors who wished to etch or carve a memento of their voyage.

Fashioning items of every conceivable variety from this material, or 'scrimshawing' as our Yankee cousins call it, is a welcome relief from the incessant repairs to tools, boats and clothes. I have seen examples of the craft change hands for more than a week's earnings and, in keen emulation of Yankee enterprise, some local whalers have lately taken to producing their own efforts.

One sailor etched our whaleship on a polished tooth, but as a novice his efforts were crude. In comparison, our cooper, who was well versed in precise work and close detail, carved a bone pastry-cutter for a sweetheart back in Hobart-town.

Such items are so far removed from the implements of bloody slaughter and gory dismembering, and so delicate and pristine is the polished white material, that few could fail to be charmed by their novelty. Yet even the final caress or polishing as this scrimshaw leaves the hand of the sailor is testimony to how hard-won is the beauty of these artefacts: in lieu of sandpaper to smooth and polish whalebone and teeth, the whaler resorts to the skin of the shark — an excellent substitute.

I had long admired the examples of scrimshaw and envied the skill which Yankee whalers displayed in this art. The pastime had become so prevalent that it was an unwritten law of the whaling world that the lower jaw of the sperm whale with its teeth should always belong to the crew to use as they pleased. Some of these men produced delicate handiwork with the crudest of tools, or even with a jackknife alone. Their design would be scratched upon the smooth surface and then colours, such as India ink or soot from the try-works, would be rubbed into the incised lines. Now, with whalebone and tools enough to occupy my anxious wait ashore, I resolved to produce something in the same line for our nursery.

My offering was completed to my satisfaction only some weeks after Henry was born early in 1843, but even then it was not bestowed until his christening at the very end of that year. Since that day my lovingly carved miniature whale became his favourite toy, eclipsing all other playthings. And later it would even display the visible marks of his infant exuberance, when he resorted to using its smooth comforting shape to relieve the pangs of teething.

DAYS ASHORE

Henry quickly became the darling of the household. His brothers and sisters, as much as his mother and myself, soon succumbed to his infant charms and in his happy ways found the best remedy for our past sorrow. His mother was determined that this new baby would not take in the grief she still felt for our lost Susannah. It would not be allowed to sour her milk.

The sad waters which had flooded our household had now ebbed and all our spirits rose in response to Henry's winning ways. This new-found happiness also manifested itself in Susannah's willingness to reanimate those social contacts which she had allowed to lapse. Virtually the only outsider who had broached Susannah's seclusion had been Mrs Cameron, and a lively correspondence had been established between the two women. One morning, after opening a letter, Susannah unexpectedly proposed a visit to the Hobart theatre once again.

It is true that Mrs Cameron had been constant in her invitations seeking to entice us back, and now it appeared that the women's long confidences together over the past year had had a secret purpose.

'My dear William,' Susannah told me with a welcome return to her former humour, 'there was more than a child in the making during this confinement. Elizabeth and I nurtured a fond enterprise designed to lift the whole family's spirits in a most novel manner.' I was quite dismayed by this, but all was revealed when she passed me a handbill that had arrived in her letter. Mrs Cameron's husband had found his local inspiration for another dramatic presentation, and had found it within our very household!

THE ROYAL VICTORIA THEATRE

Takes pride in presenting to the public
An Entirely New & Stirring Drama
In Two Acts
Specially written for a Patriotic Colonial Audience:

'The Devil & the Deep Blue Sea;
or, Young Jack Tar & the Pirates'

On MONDAY EVENING, — 1843

Will be presented, with new Scenery, Dress, Decoration, & Music

The Devil & The Deep Blue Sea

Founded upon the True & Astonishing exploits of a young Colonial boy. Abducted by a Whaling Captain, Pursued by Pirates & Bravely Rescued by the Royal Navy! Serves as a powder monkey at the celebrated Battle of Algiers.

Thereafter, covered in glory, he is sent home on a Convict Ship! The happy conclusion to this true drama finds our young Hero reunited with his family after five long years away.

Preceded by a Prologue,
written for the occasion & to be spoken by
Mrs Elizabeth Cameron

Vivat Regina! Samson Cameron, Lessee

Such was the press of friends and well-wishers, and of rivals and crewmates from the whaling fraternity, that the curtain rose half an hour late. Before this, some of my whaling companions and the boisterous crew they had brought along greeted us with three resounding cheers. All this happened after we had been finally shown to our seats with an unnecessary display of ostentation.

I doubt if some of those present had ever been inside a theatre before and wondered if the actors would receive equal notice and applause on their appearance. The house was sold out, with a noisy crowd jostling from gallery to pit. The stage-box had been reserved, not for the critics but for the Chamberlain family, and if it was not the best vantage point for viewing the stage, at least the powerful blaze of the lights enabled the audience to see us to best effect.

It was only then, when the bold action of my younger days was about to be commemorated in mimic display on the stage, that I felt as I often had before the ultimate moment on a whale hunt. And my nervous qualms were not at all quelled by the excited chorus from the family. A gentle hand in mine and a tender assurance from Susannah that I would be honoured and not humiliated by the evening's presentation calmed my apprehension. I resolved to enjoy the spectacle as it unfolded before me.

After the green curtain was rung up, the children's chatter was stifled by the small band striking up with 'God Save the Queen'. When the houselights were dimmed and the footlights opened their full glare upon the stage, I could not help remarking on the waste of whale oil. 'But, Father,' responded my eldest boy William, 'their extravagance brings grist to our mill.' The opening scene was then revealed to our expectant hush as this powerful light illuminated a glittering scene at sea. A round of hearty applause rewarded the painter for his skill and now commenced 'The Devil & the Deep Blue Sea'.

Out of the waves arose Old Neptune. Once again a round of applause rang out and greeted this tutelary genius of Old England. Thereupon a most sonorous curse was pronounced by the water god on the heads of French upstarts for daring to challenge the command and dominion of Britannia over the seas. And then the actor vanished and the water became agitated, or at least painted canvas strips representing the sea rolled back and forth. Slowly the prow of a ship hove into view whose proud flag proclaimed her to be British. My patriotic fervour, however, was sorely checked when I discerned her name: the *Frederick*.

For all the artifice and trumpery of the scene, I was at once transported back thirty years. A cold and sickly fear overcame me and I started forward as if to warn her of the approaching danger. At the very same instant a young lad sprang onto the deck and with ardent gaze and gesture directed the Captain's eyes to some distant sail. Accompanied by the ominous beat of a muffled drum, the drama was conducted in dumb-show. The Captain raised a telescope and his galvanised reaction clearly indicated the approach of an enemy ship. At this the ocean became more agitated still, to match the consternation on board the *Frederick*.

And then, slowly, the prow of the French privateer rode into view. At this fearful meeting tears sprang involuntarily from my eyes and I truly believe that only my wife's steadying hand prevented me from turning away. But then the theatre erupted into derisive calls and jeers as the French faced off to the *Frederick* and this, too, restored and roused my ardour.

Suddenly, with our expectations at fever pitch, the muffled drums reached their crescendo, then stopped. A terrible silence filled the void. The whole theatre held its breath until the immediate front of the stage erupted into violence. Strident cries rent the air as a cruel swordfight raged between the opposing crews. Into the mêlée ran the young hero of the piece, again pointing towards utmost danger at the very moment when the Captain of the *Frederick* was struck through the heart.

His agonised scream visibly affected many in our audience. Through a welter of tears I could see my own young self kneeling in a spreading circle of red. At this too realistic touch, and even while recognising its palpable fiction, I averted my eyes. Once again the muffled drums beat their sad tattoo and this time the defeated crew of the *Frederick* were led away. This was more than some in the audience could allow. Angry raised fists and loud oaths ushered the players offstage. In consternation and grief some in the audience even had to be restrained from leaping into the fray.

After the *Frederick* withdrew, an angry storm lit the backdrop and heavy drums and rattles announced thunder and lightning. Then, into this new affray, with stately bearing, came the prow of a Royal Navy frigate. All our spirits were again reversed. The prisoners below broke out of the hold and the deafening roar of a cannon announced that the British lion had been aroused.

The audience also added to the clamour, so great was their enthusiasm to urge on these champions. Behind me I could hear my own sons rallying the cast; even little Henry at so young an age was involved. The loud cries and vocal audience actually seemed to rouse his spirits rather than alarm him and Susannah needed to restrain his flailing little arms.

Once again the very front of the stage became the site of a bitter struggle and here, with ill-timed courage, right into the midst of the skirmish darted our young hero. His appearance elicited a hearty cheer of support from the audience. Susannah's hand tightened on mine in expectation of the drama's next, most telling, strike. Just when it seemed that the enemy was subdued, a sudden and vicious thrust stopped the boy as he cheered on his rescuers.

Others besides myself were now so involved in the drama that they had forgotten it was only a performance. I gave an involuntary shudder at the cowardly strike on stage but steeled myself for more. I had not served under gunfire at Algiers merely to quake in public now before its re-enactment. This first act concluded with a touching scene in which the good Dr Quarrier supported the injured child, while all on stage broke forth into a spirited rendition of 'Rule Britannia'.

Interval was a bustling crowded affair, with old friends and those only newly acquainted with the colourful exploits of my youth pressing forward to meet me. I am sure that I did not relish being their cynosure and was grateful when a drum roll announced that the second act was about to commence.

After we had regained our seats, a series of placards were revealed, each one read out in turn by a stentorian voice.

'*Scene*: The Ramparts of Algiers. The Royal Navy arrives to conquer the infidels and release Christian slaves.'

'*Players*: Admiral Exmouth leads the Fleet. William Chamberlain, now recovered, has enlisted and serves as a powder monkey.'

'*Action*: The Battle of Algiers opens with an emissary delivering the British terms to the Arabs.'

And so the drama continued. Yet for all the brave attempt on stage to re-create Algiers, a more poignant panorama was unrolling within my mind. Truly, I felt as if I were reliving the same exertions and feeling the same frenzy of thirty years before. And so the second act played out before me as my hands grasped the railing on our box like a vice.

All too quickly the battle was won and as the victors broke into another rendition of 'Rule Britannia', the whole theatre erupted into a combined chorus. I relaxed my hold on the handrail and wiped my brow. I felt as exhausted as if I had fought aboard the *Leander* once again.

THE BUN-YIP

Four years later, in 1847, I was sadly called to Sydney to attend the deathbed of my dear mother. Her remarkable health had sustained her for so long, but it was obvious to family, let alone medical opinion, that she would be unlikely to last another year. This was my first visit back to my home-town for many years and I hardly recognised the bustling metropolis it had become.

A clear indication of growth and prosperity was the publication of a much more substantial and daily newspaper than ever before. As I scanned its listings of shipping arrivals and departures, auctions, local and international news, advertisements for the painless extraction of teeth and all the accumulation of events which define a thriving commercial centre, one front-page notice aroused my interest.

THE BUN-YIP

THE PUBLIC ARE respectfully informed that by the kindness of His Honour the Speaker, the cranium of the supposed Bun-

Yip, recently discovered in the vicinity of Melbourne, may be inspected at the Australian Museum upon Wednesday and Thursday next, from ten to three o'clock.

Duly accepting the Museum's invitation I presented myself at its temporary premises in the New Court House in Darlinghurst, only to find a milling throng gathered outside, all equally curious to see this notable and bizarre denizen from Aboriginal nightmare.

The crowd jostled me along, their crude desire to be astonished and even shocked by the exhibit obvious in their straining stances and roving eyes. The gathering was so dense that even before I confronted the raree-show itself, I was ready for a cleansing draught of sea air. Imagine my disappointment when I was elbowed past a glass case containing a small skull, significant only for the extreme deformity it displayed. 'So this,' I said to myself, 'is the wonder of the day! I've heard of stranger things on my travels — freaks of nature, some of them, or even clever hoaxes created by inventive showmen.'

I was reminded of one of the seamen on the *Frederick* all those years ago, whose stories reduced the entire crew to my own credulous level. Once again we became an audience of children eager to be entertained by stories from a world of wonders, frightened by what could lurk beneath the surface of everyday life. This sailor swore by a most sacred oath that on an earlier voyage, his foray ashore into the market at an exotic port of call had yielded the most unexpected wonder of his life — a mermaid! He admitted that she had not been the seductive siren of mariners' folklore, the flaxen-haired temptress arching her glittering scales in the brine and offering her wave-covered bosom to beguiled sailors. Instead, he found a small wizened body floating inside a large glass tank. How could something so rare and fabulous be so ugly? Even as he stared at this bizarre exhibit, some jostling effect from the crowd — surely it was only that — sent a ripple through the tank and slowly and with mesmerising effect the mermaid revolved until one fishy eye fixed itself upon our spellbound narrator.

Our appreciative gasps were cut short by his raised hand. 'Don't think you for a moment that any of this was God's creation,' he warned. 'Next to me was a surgeon from a British frigate whose trained eye exposed the flaw in our mermaid. This surgeon was provoked by my credulity as much as by the skill of those who had assembled the monster. "Surely, man, you can see that this is no siren of the seas?" he said, shrugging his shoulders. "There's been some clever surgery here that has grafted a monkey's torso on to the tail of a large fish!"'

I recalled this story and its clever fakery as I peered into the glass case displaying the skull of the 'bun-yip'. I could easily tell that this was was no man-made exhibit; for all its deformity it looked no more bizarre than other natural specimens I had seen over the years. But people love to be entertained and a shiver of dread is also welcome sometimes. I left the Museum shaking my head and marvelling at the crowd still forming outside.

That evening, my attention was again claimed by the *Sydney Morning Herald*. It contained a long letter from a W. S. Macleay on the bizarre skull and its possible origin. Here was a voice of reason and rare learning — a welcome contrast to the excited and fantastic nonsense which I'd heard voiced that morning at the Museum. 'If the skull should eventually be proved to belong to a distinct species,' Macleay wrote, 'this new animal must be placed between the horse and the lama.' He went on to explain that 'the Bun-yip is said to be a solitary aquatic animal, whereas this skull must have belonged to a creature which, if full grown, would have delighted in grass, dry land and the society of its own species.'

The writer obviously drew on a wealth of learning and had the additional benefit of an unparalleled personal collection of specimens, which had honed his scientific thinking. His letter continued: 'I have in my possession a cyclopean skull, the foetus of a mare, which was found floating on the River Hawkesbury in the year 1841. In my judgement, this cranium of the supposed Bun-yip, when compared with the one from the Hawkesbury, only serves to show the extreme limits between which all monstrous variations of the place of the eyes in the horse can possibly occur.' In conclusion this learned gentleman described the skull on exhibition as a *lusus naturae* — a sport of nature.

Macleay's letter piqued my interest with its sensible and calm words which were designed to quell the enthusiasm of the Sydney public for their so-called 'bun-yip skull'. After viewing this grotesque osseous anomaly myself, I was relieved to read his erudite warning to those who wished to proclaim the skull as serious scientific evidence of a new species peculiar to Australia. In particular, he cautioned the uninformed whose zeal to parcel up the find and send it to England could only entertain the learned societies there with a display of Colonial ignorance and credulity.

When I found that Macleay was a trustee of the Australian Museum I resolved to meet him. This proved more difficult than I had at first imagined. The man was spoken of as something of a recluse, and it was only when a mutual friend commended my enthusiasm that an invitation to view the town's most renowned private collection of specimens and curiosities was offered.

Macleay had recently moved into the family's imposing residence, Elizabeth Bay House, after he had taken over his father's affairs. Here, on the side of a snug little bay and with a lawn of English meadow-like verdure in front, Macleay's father had established one of the Colony's most desirable urban estates. Originally, fifty acres of land had been granted to this former Colonial Secretary, although subdivisions had since reduced the size of the property. The garden, which had been laid out in the 1830s, was even known back in England. It was regarded there as a notable antipodean effort and one which had sent valuable specimens home to Kew Gardens. Although no gardener myself, I was still astonished at the profuse cultivation of native species, particularly the ferns, along with exotics. Here was a picturesque haven which neutralised much of the impact of the deportation element in Sydney society.

William Macleay, the son, now lived there in a solitary but not isolated style. To the friends who visited, he poured forth with a memory quite astonishing the varied and extensive stores of an encyclopaedic mind. Nor did I find him the recluse that popular report had suggested. It is true that he was naturally reserved and found recreation and solace in his extensive library, in the unique family collections now in his care and in his gardening and the culture of orchids. The tropical climate of Cuba, where he had resided for some time before his arrival in Sydney, seemed to have prematurely aged him. He was further burdened with gout, a malady which none find conducive to amiability.

Yet he was no misanthrope or hermit. Here was a man who had met Darwin in England and had encouraged him to publish a 'Zoology' of the voyage of the *Beagle*. Even so, he still found much of interest to discuss with an old whaling captain. We carried on an animated conversation and took a frank and spirited delight in each other's company and opinions.

The Macleay family vision of founding a permanent home for the Australian Museum, in a building dedicated to its display, was already being realised. While the initial collection was temporarily housed in Sydney's New Court House, construction was slowly progressing on an imposing sandstone building closer to town.

I had much to tell my family after I returned to Tasmania. 'What I witnessed there, at the Australian Museum,' I explained later to Henry, 'was a significant example of ignorant enthusiasm.

Remember, my boy, that there are two things which most enlarge the intellect — travel and reading — and it is obvious that many of these Colonists are stay-at-home blockheads.'

Yet, in all fairness, I could also concede that my own countrymen were no more or less credulous than their English cousins. I recalled a passage in a celebrated set of volumes edited by the esteemed British authority Sir William Jardine only a decade earlier. One individual book in his *Naturalists' Library* was entirely devoted to whales and discoursed with admirable authority and expertise upon its subject. Its illustrations as well as its text had so beguiled me that eventually I procured a copy. I still keep it among my favourite books. Yet even in its erudite pages I find errors, let alone colourful claims that amuse me still.

Surprisingly, I found that the more bizarre denizens of the mighty deep were accorded credit and space by this respected man of science. Within his lengthy introduction he thus proposed a companion volume with a fabulous promise. 'It is proper to remind our subscribers,' he wrote, 'that a portion of the great order "Whales" still remains undescribed. It is therefore proposed ere long to devote a second volume in which we shall also endeavour to trace the reality and history of a race of huge animals, known under the popular titles of Mermaids, Sea Serpents, &c., and which, while we cannot doubt their existence as singular creatures of some kind, we are totally at a loss what form or nature to attribute to them.'

THE HISTORY OF THE CETACEA

Some few years later, a most unexpected and singular gift arrived at my door, delivered by a visitor from Sydney. It was a pamphlet dated 1851 and bearing an intriguing title: 'History and Description of the Skeleton of a New Sperm Whale, lately set up in the Australian Museum.' The pamphlet arrived together with a letter from my acquaintance, Mr William Sharp Macleay.

It was with a fond memory of this entertaining and hospitable gentleman that I read the letter, inserted into which was another bonus: cuttings from the *Sydney Morning Herald*. One from late 1849 announced that the carcass of a sperm whale had been found at sea and towed into Sydney Harbour. After two months on Pinchgut Island under a preparation of lime and other chemicals, a perfect assemblage of bones was obtained. The skeleton, just over thirty-three feet long, was then painstakingly assembled and finally exhibited along the outside wall of the Australian Museum, opposite Hyde Park. Here it was a daily spectacle, competing for attention with the building still being constructed.

My delight in this novel addition to Sydney's Colonial attractions was considerable, but I was hardly surprised. I recalled my conversation with Mr Macleay five years earlier. 'The great evil in these Colonies,' he'd complained, 'is the absence of scientific men. Many of the settlers here received some education, but there are few who can fairly be called men of science and the consequence is that the half-educated, with but a smattering of knowledge, are able to lead the more ignorant by the nose.' Upon showing me some curious animal bones from his collection, including a cyclops skull which was easily a match for Sydney's 'bun-yip', he had then solicited my opinion as to a future museum exhibit. 'Would not a full-size whale's skeleton, intricately assembled and displayed, be a dramatic and splendid eye-opener to these uninformed Colonists? Instead of fawning over a deformed horse's skull,' he said, 'I believe that a much more majestic skeleton should take pride of place in our new Museum.'

Here, indeed, was a singular idea and one well calculated to appeal to my own interests. My response was immediate. 'While a lion or a tiger has become quite a vulgar animal in our menageries, there are few persons who have ever seen a live cetacean. Furthermore, perfect skeletons must be very rare in museums.'

Now, some few years after I expressed this opinion, I found these very words of mine echoed in Macleay's pamphlet. I could not doubt that my own passion had enthused him to broadcast his desire for just such a magnificent specimen.

While others might have regarded this as an outlandish suggestion, he had seemed to relish it immediately. 'Whales are such enormous creatures that the ordinary methods used in the study of other animals cannot be applied to them. Instead of having actual specimens before him for comparison, a naturalist must depend almost entirely on notes, measurements and drawings,' he lamented. 'Thousands of mariners have captured and cut up whales and have never accurately examined one of them; and yet it is upon their vague descriptions and figures that zoologists have endeavoured to establish the natural history of the cetacea. This is the true cause why the history of the whale is so meagre, yet so full of contradiction and repetition.'

Perhaps somewhat arrogantly I responded with a question. 'How could lubbers ashore be expected to know the real facts of deep water when they squat here and are beguiled by the sea-shells and other waifs which the sea tosses them? A whale floats ashore and they make drawings that could stand as well for a bare hillock and put it in their picture-books as one of the living facts of God's creation.'

My host responded with a gentleman's indulgence towards my unthinking criticism. 'For whalers like you, gain is the only view and the profits of a nauralist like me are considered as secondary points, if considered at all. At the best, our opportunities for examining such animals do not usually occur till the parts are in such a state as to defeat the purposes of accurate enquiry. And even these occasions are so rare as to prevent our being able to imagine, by a second dissection after your slaughter, what has been garnered in a first.'

Warming to his subject even more, my host became quite animated with enthusiasm. 'For all our supposed civilised knowledge, Captain Chamberlain, the ancient Greeks were in awe of these creatures.' His keen eyes sought mine and as he observed my eager interest, he continued. 'It is virtually forgotten knowledge today, but over three centuries before Christ, the great Aristotle wrote about the smaller cousin of the whale. He believed that dolphins communicated with each other by their murmuring sounds.' Here he regarded me quizzically. 'But it may be that you think any talk of such animal attributes too fanciful for a man of science?'

I dismissed his question and assured him that I had seen things on my voyages that would repay his scientific scrutiny and enlighten the stay-at-home public. 'In my experience, sir, I have observed that whales clearly have a means of communicating with each other at long distances; certainly at distances as great as are commanded by the eye from the masthead of a ship. The means are a mystery, but every whaler knows it for a fact and bases his operations in the chase upon it.' This assurance of mine that much knowledge remained for scientists like my host to uncover encouraged Mr Macleay to voice further opinions of his own.

'It has been suggested that, as water is a good conductor, these creatures may communicate by sound. But I would counter that the distances are too great for any sound which the whale is capable of making to penetrate. Furthermore, it has been observed that this telegraph is perfect as ever in high winds when a thousand waves are breaking. Can you explain this, Captain Chamberlain?'

I could only shake my head in perplexity. 'Dart an iron into a bull whale or gally him by going on his eye and almost simultaneously as he raises his cutting flukes into the air, the whole school will show alarm. They will disappear from the surface and only come up miles to windward and run head out. If it be a cow that is struck, the bulls are arrested in flight and are apt to gather about her. Again, when a school of cows and calves are running frightened to windward and a calf be struck, the whole school will bring to and gather closely around the wounded young. Sometimes, indeed, so closely packed are they that the enclosed boat would not dare to use the lance. But should the iron draw or the calf die, the whole school will instantly scatter.

'Even more to your point, sir,' I continued, 'I have observed the effect on schools miles to windward when a boat is going on to a whale to leeward. As soon as your eye could turn from one spot to the other, the alarm of a struck whale to leeward would be communicated to those to windward. These are facts within the experience of all whalemen, although the manner of communication is not even guessed at.'

And now, in his pamphlet, I had a tangible example of his enquiring intellect before me. However, not surprisingly, I found that much of the scientific detail presented was well beyond the average Colonist's capacity to appreciate. Even so, it occurred to me that the learning displayed here marked an extraordinary advance in scientific explication for a Colony which, only four years before, was still chasing chimeras like the bun-yip. Whatever the censorious may once have said and thought of Sydney learning and Botany Bay science, this work ought, for all time to come, prevent their sarcasms having any influence. In my letter thanking Macleay for his gift, I could not resist a touch of levity, which I had no doubt would be appreciated.

> I marvel at how a pamphlet as small as this can accommodate all the characteristics and affinities of the largest of Earth's creatures. Your hand, sir, is surely evident in this anonymous work, to which anyone should be proud to append their name. It contains wit and wisdom, as well as blubber and whalebone, and an array of scientific etymology which must have frightened its compositor out of his senses.

THE WONDERS OF EDEN

Over the following few days, I found my thoughts returning again and again to the idea of that singular skeleton on display outside the Australian Museum in Sydney. Finally, I resolved to visit the town once again and to renew my acquaintance with Mr Macleay. This time, however, I would not travel alone; I would take my youngest son, Henry, as a companion on my journey. Of all my children he was the dearest to my heart; his birth an unexpected late arrival which touched an old father's fondest affections.

I was keenly aware that I hardly knew this youngest son of mine and I welcomed this as a rare opportunity to spend time with my growing boy. Henry reminded me so much of myself at his age — there was a dreaming reverie to which he so often succumbed. In contrast, however, I felt there was a disinclination on his part to go whaling like his older brothers. And so I planned to take a long and overdue break from my years of whaling. With the older boys already experienced and eager enough to maintain the Chamberlain family presence in the fishery, Henry and I took passage on a coastal trader going to Sydney.

One attraction in our itinerary would be afforded by the vessel's calling in at Twofold Bay. It had long been a keen desire of mine to visit the whaling town of Eden halfway between Sydney and Port Phillip. In 1828, the year before I married, an item in the *Sydney Gazette* had caught my attention by its prominent mention of Captain Raine. That man had been a hero of sorts to me ever since my meeting with him in 1821. On that occasion, he had added further acclaim to a name already illustrious when he sailed into Sydney with three survivors from the whaleship *Essex*. And since then I had heard the most astonishing stories of killer whales herding their larger cousins into Twofold Bay for Raine's whalers to capture — almost like reliable sheep dogs who harry the flock into their pens.

Such is the bay's situation, too, that it affords unparalleled shelter from winds on the whole coast south of Sydney. It is therefore most valuable as a place of refuge, as well as providing a great facility for ingress and egress. Unfortunately, none of the settlers, including Captain Raine, remained in the original township of Boyd-town. Its founder, the empire-building Benjamin Boyd, bolted for the Californian goldfields then sailed off into oblivion two or three years before our visit. It was widely known that he left behind the most staggering debts and his abrupt departure threw a pall over the town's name.

Even so, the bay still advertised its attractions in our Hobart newspaper. Our passage north would include a stop there to refresh and take on wood, water and provisions. And since all these services would be free of port charges and pilotage, the bay may eventually regain its good name.

When I identified myself to the ship's master as a fellow sea captain, he invited us to go aft as often as we liked, adding: 'We've a gentle voyage ahead if this fine weather holds.' Looking down at Henry, he laughingly offered him a place as a cabin boy.

'Thank you, but we're going to Sydney — we're going there to see the skeleton of a whale!' Henry enthused. 'I'm going to write an account of it in my journal.' He was eagerly clutching an old logbook of mine whose blank pages were being filled each evening with the day's sights.

And now our Captain revealed that he was acquainted with my personal history. In the small realm of society that exists in Colonial Tasmania, it appeared my early adventures still retained a popular currency. 'I am honoured to have you aboard, Captain Chamberlain,' the master went on, 'although I expect our short run up the coast cannot compete with a whaler's adventures — or, indeed, if there is any truth in the telling, with your hazardous exploits as a boy.'

I assured him that the only adventures we anticipated on this trip were such as might occur during our brief stay in Twofold Bay. 'Is it true,' I asked, 'what I hear of whaling there?'

'Well, Captain Chamberlain, that all depends on what you hear. The scandalous leave-taking of Mr Boyd is still recent enough for much ill-will to be felt towards the place. People prefer to talk about Eden now, rather than honour that miscreant's name by mentioning Boyd-town. But I'll admit that the fellow did have a vision.'

I would not begrudge a man his dream, so long as it was not at the expense of others. For myself, I see long hard years of voyaging fulfilling my fondest dreams as my grown sons follow their father's example. But Boyd's empire-building was infinitely more grandiose, as my companion revealed.

'Boyd over-reached himself and fell from grace. He tried to unite two frontiers, the inland with the oceanic, but left the management in the hands of an inept crew. His namesake is now fast becoming a deserted town while its rival, the settlement of Eden across the bay, has won Government support. And yet even this may not save the bay.' He spoke regretfully.

I recalled his pessimism three days later as we sailed into Twofold Bay. Our arrival was heralded by the sight of a tower on the south head. Even twenty-five miles out to sea, its white Sydney sandstone caught the light. Then, as we got closer in, the spire of a handsome Gothic church vied for this heaven-reaching distinction. Ben Boyd's crenellated lookout tower stood sentinel on the southern promontory of the bay, which was roughly in the shape of a 'W' and overshadowed by the dark mountain flanks in the distance. Deep in the centre of this bay, a small hilly point directed our eyes to the township of Eden.

In response to my delighted exclamation as to the bay's natural advantages, the Captain made his most pertinent observation yet. 'It is quite true,' he said, 'that the town's principal asset is its easy access by sea. But this is compromised by its otherwise complete geographical isolation. While its chief products have included blubber and wool, its vulnerable location has seen other, less envisaged, exports.'

And here the Captain showed another side to his erstwhile serious discussion. 'Bolting wives fleeing from isolation and loneliness have been a major and unforeseen export. And it seems their example set a precedent for the absconding Mr Boyd himself.'

I could see that Henry was much delighted with the prospect of anchoring in Snug Cove on the north side of the bay; true to its name it offered a sheltered mooring. Among the first to hail us from the shore was a small group of Aborigines. This was another novel experience for Henry. Back in Tasmania the Native population had been in decline for some years and, because it was almost exclusively confined to a designated reserve, there was virtually no intermingling of the races. But these fellows seemed quite at ease in the whaling town.

In response to my questions, we were told that Boyd had assembled a crew of whalers from these Natives. In their various employments he had found that the Aborigines contradicted the hasty conclusions which so many superficial writers had drawn in reference to their faculties. Our Captain's knowledge of these crews produced even more startling information.

'Among the Aborigines in this area,' he explained, 'there is the peculiar belief that, after their death, certain honoured members of the tribe can be reborn as killer whales in this bay. In their own idiom it is regarded as a reward to 'jump up' in this new guise. So firm is this belief that they even particularise and identify certain individual killer spirits. Just how this belief first developed I cannot say, but it is certainly as colourful as any other story you will hear here.'

Once ashore, Henry and I were given directions to Alexander Davidson's house. Months before, I had written to this associate of Ben Boyd's and proposed a brief visit to Twofold Bay. I was delighted when this complete stranger had responded warmly and, with fraternal hospitality, invited us to stay with him for an overnight gam. Our host had such stories to tell us that night that Henry insisted on keeping me company far beyond his normal bedtime.

In May of each year, the increasing cold of the waters in the Antarctic Ocean causes the whales to turn towards the Northern waters in search of a warmer and more congenial clime. On their return four or five months later, after the cows have given birth, they sometimes cruise into Twofold Bay. Here they have been observed rubbing against the rocks in an endeavour to rid themselves of the many barnacles which adhere to their bodies.

'One hears many preposterous and problematical stories about the doings of the whales, but I assure you,' said Mr Davidson, 'that what I will tell you is from my own personal experience.'

After whales are sighted from the lookout, the men burn bracken fern to create a smoke signal and the whaleboats are launched. In the meantime, however, and this is where fact begins to sound like fiction, a second message arrives, not from the land but from the sea. As vigilant as the lookout men might be, their attempts are as nothing when compared with those of the killer whales. These harriers of the deep possess an organisation and an intellect that far exceeds the strength and bulk of their migrating cousins.

'I have known as many as eight whales captured in one single day off Twofold Bay,' Davidson proudly informed us, 'and all were achieved with the vital assistance of these killers herding them into the bay. They throw themselves out of the water, floptailing to alert us to the quarry, and then the cry of 'Rushoo!' will resound around the bay as we scramble for the boats. In this remarkable manner, the whale is harried into the bay and in the distraction we are enabled to plant our harpoons into it. When we arrive, the killers will erupt into frenzied excitement, leaping out of the water and racing from boat to boat like enthusiastic terriers. They attack like a pack of dogs, leaping onto a whale's back and fastening to its lips to prevent it sounding. Should it escape this first onslaught and attempt to make to sea, it finds a second cordon of killers waiting. After they finally take the whale down bodily, we leave a buoy to mark the place where, some days later, the inflated carcass will appear.'

Davidson smiled at us. 'Then you would witness one of the little amenities and courtesies which are common on both sides between us and our pack of killers. For now it is found that they will have eaten the whale's tongue — an organ weighing, as you would well know, up to two thousand pounds. No doubt this delicacy is regarded by them as their perquisite. It is also true that although these killers will attack a whale, yet they seem peacefully disposed towards mankind. Should a party of whalers upset their boat, then these killers do not attempt to attack them.

'Eden has seen all this and Eden no more discredits the story than a bishop would discredit the miracle of the loaves and the fishes.' And this was Mr Davidson's solemn and final assurance.

But it occurred to me that the killer stories implicitly believed around Twofold Bay might place too much strain upon the credulity of Australians elsewhere. For example, our host related that on one occasion he was being beaten in the chase by his opposition. Several of the killers kept swimming back as if to urge him on, while others were lashing the water in front of the opposition, apparently in order to impede their boat. Knowing the habits of these killers and wishing to test their allegiance still further, Davidson threw a line over from the bow of his boat.

'Immediately,' he declared, 'it was grasped by two killers who started for the quarry. Our boat was taken along at such a pace that we soon passed the opposition and got fast to the whale.'

And more in this vein was to follow. 'On another occasion,' Davidson recalled, 'I was steering when a sharp turn of the whale, and hence of our tethered boat, landed me in the water. The boat being fast to its prey, the crew could not stop to pick me up. But truly, all the time I was in the water two killers swam round, yet never once attempted to touch me. I had no fear of sharks either, not with these guardians. And immediately our relief boat arrived to rescue me and I was safe, the killers headed for the whale.'

Not surprisingly, the whalers have given fond names to some of these killers and believe they have a special affinity with them. However, it is my opinion that Davidson's yarns have been repeated so often and amplified to such an extent, and are told with such simplicity, that a stranger feels almost afraid to disbelieve. At any rate, any disbelief must be hidden — if a listener once cavils at a story he is told no more. One anecdote, however, was simply too much to swallow and has afforded me great amusement ever since.

We were introduced to an acquaintance of Davidson's, an old whaler who was somewhat deformed. He appeared to be regarded with something approaching veneration by his listeners, who looked upon him as a sort of hero. It had happened that he was steering a boat in the chase after a very large whale which kept attacking them. The steer-oar had been broken when the whale charged once again, open-mouthed. Our hero now faced the whale with both feet hanging over the stern, the oar-cleat firmly grasped in his hands. Just as the whale was about to smash the boat to kindling, he dashed out both feet and hammered his iron-heeled boots into the animal's forehead. The blow was well delivered and the impact shot the boat many feet clear of the whale, which abandoned its pursuit in shocked dismay and swam off.

And then, with great pride, our hero displayed the proof of his story. The sinews of this old fellow's legs were so injured that he had been bandy-legged ever since. Truly, people who can believe this yarn are of great faith!

But there was more. A final wonder was revealed to us during our visit and, even though I counted myself a seasoned whaler, I could scarcely credit that people would subject themselves to what was described. With many visitors desperate for relief from ailments like rheumatism, it had lately become the practice for such sufferers to resort to a novel remedy. A hole was dug into the carcass of a whale and the patient encased in it up to the neck for as long as he could withstand the considerable heat and the rotten stench. It was claimed that the whale acted like a huge poultice, opening the pores of the skin and imparting unique benefits to the afflicted.

While long-term benefits might be questionable, the short-term result was quite guaranteed — for a week or more afterwards the patient gave off a dreadful odour, abhorrent to man and beast.

INSIDE THE WHALE

Once we reached Sydney I discovered that Henry's enthusiasm over the whale's skeleton was not matched by mine. While I could indulge a father's fond smile over his awestruck reaction, this was not the prey I had hunted. Stripped down to its bare assemblage like this, it had lost, to me, all the majesty of its mighty upheavings. This proud Leviathan had been reduced to a stranded exhibit. Why, they had even painted its bones with the same paint used on the surrounding fence!

I had imagined that the Sydney public had now moved beyond the credulity and ignorance shown only four years before, when they gaped at their 'Cranium of the Bun-yip'. Now I realised that this was yet another sideshow, one which gave no idea of the whale in all his dreadful might. With man, the skeleton is the mould of his invested form, but with the whale this is not so. The great head and heaving body that enfolded these bones remained unimaginable from his skeleton; they were beyond any dull-witted landsman. Only in the threat of quickest annihilation, only within the eddyings of his angry flukes, only when borne on the profound unbounded sea, could the fully invested whale be truly and most awesomely discerned.

But I would not scoff at Henry's enthusiasm. This was his first adventurous journey away from home, where the most freedom he was permitted was his daily walk through the bush to and from school. Now, in our passage north to Sydney, he had shown himself to be a stalwart sailor. He reminded me so much of myself at his age: he was, after all, only a year older than I had been in 1811, on my own first voyage. But what a distance in time and anguish separated those two experiences of father and son! Our shared journey now brought us even closer together and his excitement at a complete change of scenery was invigorating. I saw his whole life stretching before him as a great adventure. I prayed that he would never have to cross the same stormy seas that I had experienced at his age. I was grateful that I had been spared to know the joys of fatherhood and, as I fondly smoothed Henry's hair, I believed that he would always be this secure and carefree.

I can still recall, with such an aching loss, the look of pride in Henry's eyes. His hand took mine as he sought out a friendly audience in the milling crowd, eager to tell anyone who would listen that his father here had wrestled giants like this and won the day. Emboldened by his boasting, he then showed an entirely unexpected show of bravado. While I was momentarily distracted, he slipped away and ventured right inside the giant ribcage.

'See me, Father,' he cried, 'I am like Jonah waiting for the whale to spit me out!'

I turned in some alarm, fearful that Henry might be in danger or even that he might damage the giant skeleton. But Henry's engaging pose and the amused smiles of onlookers wiped any foreboding from my mind. I gently chided the boy.

'Henry! Enough. Climb out and be more respectful. That is no place for play. The whale is God's mightiest creature in all the sea and you must not mock the Bible's message.'

Henry swaggered back to join me. This bitter-sweet image returns vividly to my mind today. I see myself reaching out, as I so often had before, to clasp him to my side. I know he was surprised at the warmth of my embrace and perhaps even a little embarrassed. But if I could have seen the dark horizon that loomed over us then, I swear that I would never have relinquished that embrace.

Then, in the midst of this fond moment, I was hailed by name and turned to receive the tremulous handshake of Mr Macleay. That this august scientist should remember me was indeed flattering and our conversation was as easy as if it had been only yesterday when we parted. But I must admit that my thoughts were only half with my acquaintance. While he steadied himself on my willing elbow, an occasional darting glance to the side discovered that Henry had again been tempted to venture inside the whale.

There, in that cavern of bones, I knew he was imagining himself leagues out to sea, diving to the greatest depths and then sounding above with a tremendous spout. Later that evening I knew he would confide all of this to the journal in which he so zealously documented his small joys, his daring exploits and his dreams of glory.

I remember, too, that this time Henry was coaxed out of the whale not by my stern gaze but by an imposing recitation from Mr Macleay.

There Leviathan,
Hugest of living creatures, on the deep
Stretch'd like a promontory sleeps or swims,
And seems a moving land, and at his gills
Draws in, and at his trunk spouts out a sea.

These memorable lines captured Henry's interest and he returned to my side as our orator asked if we were familiar with *Paradise Lost*. I knew that when this normally retiring man displayed a vocal pride like this in the Museum — in his Museum — it was a singular accolade to my presence and to our shared enthusiasm.

These carefree days still remain my fondest memories of time spent with my dear son. But even then there was, unknown to us and massing in the darkness, the slowly gathering momentum of an approaching wave that would tear us apart.

THE DEVIL ON LAND

HOBART TOWN DAILY COURIER

[Friday] 14th November 1856

From the information which has at present reached town, it seems that Henry, the unfortunate son of Captain Chamberlayne, of Blackman's Bay, a boy of about 14, left his parents' house about 8 o'clock on Wednesday morning, for his school at Mr Oldfield's, Kingston. The poor lad not returning at his usual time in the afternoon, a messenger was despatched to Mr Oldfield's to enquire the cause of the delay in the boy's return. Upon the return of the messenger, it was made known that the boy had not reached school that day, and the agitated friends immediately instituted a search for him, more particularly confining the search to the beach, as the boy had not been seen in the township in the day, and it was supposed that he might have gone fishing, and had unhappily been drowned. This painful search was continued until Thursday morning without success, when the loss of the child being more extensively known, the neighbours of Captain Chamberlayne, with their men, lent further assistance. The search was then recommenced, covering this time the bush over which the boy usually passed in going to school . . .

[*William Cooper, an employee of Captain Chamberlain, found the boy's body the day after his disappearance.*]

HOBART TOWN DAILY COURIER

18th November 1856

I found him lying on his face. I called out 'Henry' three times. I lifted his head, and found it quite cold, with blood on his face, and I left him immediately; one foot was lying on the other. I rode away to give notice to the police, and meeting three men I left them in charge of the body, and rode home to Captain Chamberlayne and told him that I had found his son . . .

[*The Coroner was immediately called and selected seven men as a Jury. Together they viewed the body and, that same evening, George Nixon was apprehended by two detective constables. Nixon was a ticket-of-leave convict working on a neighbouring property who had been seen in the company of the young boy.*]

HOBART TOWN DAILY COURIER

17th November 1856

The body appeared frightfully mutilated about the throat, the wound commencing from the right ear across the wind-pipe, and gave evident marks of the perpetration of a most diabolical offence having been committed . . . While the Jury were viewing the body, the Coroner had the prisoner Nixon brought from the lock-up to the room where the body lay, covered for the moment: Nixon was placed close by the body, the covers lifted up, but Nixon did not exhibit the least symptom of fear or surprise . . .

[*On Monday, 17th November, the Coroner and Jury visited the scene of the murder.*]

Whoever the murderer may be, he must have enticed his victim from the track, as he was about crossing the creek, to a considerable distance up the gully, away from observation. The footsteps are plainly to be traced to the fatal spot. There the first blow must have been given, and there the foul deed perpetrated, as blood is discernible both on the ground and on the trees; but the body of the unfortunate youth was found at a short distance from where the struggle for life took place. At that spot the ground evidences a desperate conflict. To where the murderer took the body – the head lying on a tussock of grass, the face downwards – there are marks of some considerable depth in the ground of a knee and toe, indicating the position of the monster as he knelt over the body of his victim; whilst the scrub bears corresponding marks of his feet, as the bark is abraded in many places.

[*Among the graphic details of the crime disclosed in the Tasmanian papers from mid-November 1856, there was one final outrage. In the circumlocution of the day, there were hints that the murderer had first committed an 'unmentionable crime' on the boy before the barbarous act of murder. The Coroner soon discovered that 'not only had murder been committed but a crime still more revolting had been perpetrated'.*

Such a frenzied attack upon a young boy and on one who was the son of an esteemed Colonist shook the very fabric of Tasmanian society. To this day it remains one of the most gruesome and shameful murders in Tasmania's Colonial past.]

HOBART TOWN MERCURY

17th November 1856

The extreme youth and innocence of the victim, combined with the fiendish atrocity of the crime – have created a sensation of horror in the community, the intensity of which we have never previously witnessed. We can picture to ourselves the poor little fellow on his way to school – conning over in his mind his lessons of the day, or whistling, happy in his innocence, and in the prospect of boyish sports in the future; his mind free from care and suspicion, having no enemy nor cause of fear – we can picture him, we say, struck down in such a moment of youthful buoyancy and happiness.

[*On the day of the boy's funeral, the Coroner and Jury visited the scene of the murder together with the suspected murderer.*]

When taken into custody one of the sleeves of his shirt was wet – having been washed – but spots of blood were still discernible, and plainly visible on the other sleeve. His face and hands too, are very much lascerated [*sic*], bits of skin having been torn off, which was not the case when he went out in the morning. Nor is this all. He was away from his employment till 3 o'clock in the afternoon, and alleges as a reason, that he had been seeking for horses, which, he must have known, were in the stable at the time. A knife was seen that morning by his bedside. It cannot now be found; nor will the man give any satisfactory account regarding it. That a revolting offence was committed on the poor boy is placed beyond a doubt by the medical evidence; and, on the same testimony, it is equally established that the prisoner was a party to that or an exactly similar crime.

[*On 20th November 1856 the Coroner's Jury returned a verdict of wilful murder against George Nixon, perpetrated under the most revolting circumstances. The isolated township of Kingston became a whole community in mourning. Windows were darkened in respect as the funeral cortege passed, although no tribute was more poignant than the massed presence of school friends of the dead boy. They followed his small coffin in the mourning coach to its final resting place in the St David's burial ground.*]

HOBART TOWN DAILY COURIER

21st November 1856

Immediately after the verdict of the Jury yesterday the prisoner Nixon ... was placed in a cab, accompanied by an armed escort of the police, under the charge of D. C. Beresford, and brought to Hobart Town Gaol, from which prison he was shortly afterwards removed to the Penitentiary ...

[*The trial of George Nixon opened in Hobart before His Honour, Sir Valentine Fleming, Chief Justice, on 5th February 1857 and the prisoner pleaded 'Not Guilty'. Because of the nature of the evidence to be given – revealing both the brutal facts of the murder and its sexual basis – women and children were ordered out of the Court before the prosecution opened its case.*]

HOBART TOWN MERCURY

6th February 1857

The Attorney-General, now, addressed the Jury, with a view of placing before them the leading features of the case. The circumstances connected with it, he observed, had caused so much personal excitement throughout the Colony, were so notorious, and had been so often discussed, that the Jury could not deem it improper on his part, if, before he opened the evidence to them, he should remind them, and he did so with much respect, of the obvious and serious duty, which they were now called upon to perform, and of the frame of mind in which they should approach the case. Everyone of them, perhaps, had heard of, and may have formed an opinion upon it, and, it might be, had imbibed a prejudice against the prisoner, as the supposed perpetrator of a crime so horrible; they might have read, also, the comments which had appeared in some of the newspapers, in reference to the alleged conduct of the accused; but they would recollect, that they were, now, not ordinary members of the community, but were called upon to conscientiously discharge their duty towards the prisoner at the bar, towards the public, and towards their God, in accordance with the solemn oath, which they had taken on the Holy Gospels. Having noticed the comments of a portion of the Press, he felt that he should shrink from the performance of his duty, if he did not openly here condemn the highly improper tone adopted by the writers, as utterly subversive of all justice, and the fair and impartial administration of the law, for if the course recommended by those journals was carried out, no man could expect a fair trial, his case being emphatically prejudged.

[*An elaborate plan of the suburbs of Kingston, which had been prepared for the Court by a surveyor, exhibited the fatal locality. One of the most telling statements from a witness was now delivered: 'The deceased was acquainted with the prisoner . . . they used to talk and joke together; the prisoner, therefore, was cognisant of the poor boy's habits.' On 8th February George Nixon was again placed at the bar, and the Jury having answered to their names, the trial was resumed. Mr Macdowell defended George Nixon.*]

HOBART TOWN MERCURY

5th February 1857

. . . the case not only concerned the prisoner, it concerned the whole community, for if the crime with which the prisoner was charged was committed with impunity or without protection, no child would be safe from brutal violence ... If the mourners who had followed young Chamberlayne to the grave, had met the prisoner, surrounded though he were by a host of constables, there would have been no necessity for a trial this day, as he would have been literally torn to pieces ... Mr Macdowell always believed, that in proportion to the enormity of the crime, should be the positive establishment of its proof, and in the present case he called upon the Jury to be very careful indeed in considering the evidence which had been laid before them ... Would the Jury then announce the prisoner guilty of such a crime upon the facts which had been laid before them? He could not conceive it possible ... He called upon them under the solemn and awful obligation of their oath, fairly, impartially, steadily, and according to the evidence, to try this case as between the prisoner of the bar, and the Crown, and to render that justice, which they were soon to administer: he had not addressed himself to their passions, but to their reason ... He did not ask for a merciful verdict, God forbid he should in such a fearful case as this, but he demanded a just acquittal; he called upon them, in the name of God, in whose presence they then were, to dismiss from their minds, all that they had heard out of court, and to take care, that by their verdict an innocent man should not be made to suffer ...

[*Apart from this eloquent plea, Macdowell had nothing else to bring before the Court. No witnesses were called for the defence. The Jury retired at twenty past five in the evening and returned at seven o'clock with a verdict of guilty.*]

A stifled murmur passed through the densely crowded Court and then all was still. On [Nixon] being asked what he had to say why sentence should not be passed upon him according to the law, he replied – 'I do not wish to shew any bravado but I am innocent. That is all I say.'

His Honour then passed sentence of death and dissection upon the prisoner in one of the most feeling addresses we ever remember having heard. He spoke of the unparalleled atrocity of the crime; of the youth of the victim, and alluded to other circumstances connected with the case which gave so inhuman and so barbarous a character to it ... He urged the prisoner to prepare for death, and to seek for mercy ...

The prisoner cried during the address of His Honour; and the strong nerve which he had possessed abandoned him. At times it seemed as if a fit of ague had seized him for the moment. Thousands of people had congregated around the Court; and a passage had to be made through them to enable the prisoner to pass to the gaol. Now that justice had overtaken them, all the feeling which had been previously shown against him disappeared; and to the honor of human nature be it said, no groan or other mark of hatred escaped any in that vast and anxious crowd.

VOYAGE'S END

The outpouring of community grief and anger and fear notwithstanding, there was one mourner who was noticeably absent from the funeral. Henry's poor mother was so affected that a newspaper reported 'serious apprehensions of her surviving the shock'. And I admit now that I was not the stalwart partner and staunch example that she most needed. For some time I was utterly and helplessly adrift on the most frenzied sea of my life.

Our own child, our youngest, stalked and murdered. And murdered only after this monster had committed a depraved attack upon the innocent boy. I would rather have found Henry's drowned body washed up on our shore as we first imagined than face this horror. But William Cooper had arrived in the afternoon, sobbing at our door. I needed but one look at his stricken face to confirm my worst fears. Instantly I was plunged into the deepest seas of anguish. My dear wife, with a heart-breaking scream, swooned and fell to the floor.

Later, after the Coroner and Jury had viewed the body, it was solemnly carried home. Against all entreaties from family and neighbours I had to see Henry's face — that sweet trusting boy. But, oh God, the too cruel marks of murder and battery that he bore! His hands, his face so pale and then, most cruel and savage, his throat. And the blood, the life's blood that had poured from that terrible wound. I was reminded then of how much blood I had spilled from countless whales over a lifetime of hunting; blood enough to still the waters, blood enough to turn the green seas red. And now our own table was like unto the deck of my whaling ship, itself stained crimson and carrying a cargo of bloody slaughter.

I reached out, as I had so often in the past with a father's automatic gesture, to smooth Henry's ruffled hair and felt my very reason unmooring itself and sliding away into frenzy. Truly, love is the sharpest lance of all and now I had been pierced through the very heart.

I bathed the body as if Henry were a baby again, with infinite tenderness. And all the while I spoke soft words that not even the rest of the family or the others present were meant to hear. I spread the small hands on mine and washed each finger, one at a time; cradled his head and took special care not to get soapy water into his eyes. And then, as I stroked his hair, I leaned in close to inhale, for the very last time, the scent of my lost child now cleansed of all violation. And throughout all this my tears fell unchecked. I immersed myself in this ritual. There was nothing and no one else involved in my intimate farewell. My whole world was centred in this farewelling and no one else and nothing else existed for me.

The violence of Henry's wounds stood out in awful contrast to my tender ministrations. Finally, the neighbours could not bear to watch our grief any longer. Silently they crept from the room and left.

Time passed.

Eventually I was able to raise my head and go outside. No one moved or spoke. There was yet one last sacrament to perform. I returned with a bolt of sailcloth and with infinite care wrapped Henry's young body. I had resolved to keep vigil there throughout the night and had swathed Henry in a treasured piece of cloth left from the first whaleship ever built in Hobart.

Now there was nothing more to be done, no further preparations for the morrow. Mercifully, my numb sorrow finally delivered me into a deep sleep, from which I awoke still clutching Henry's bloodless hand.

17TH NOVEMBER 1856

Today we delivered Henry to the family vault at St David's. Many friends and even complete strangers came to offer condolences. The death of my son touched so many and all had words of gentle, if futile, wisdom. They talked of 'tomorrow' and the 'future'. Although they meant well, a voice inside me could only add, to each and any thought of days ahead, two words: 'without him'. Tomorrow 'without him'. The future 'without him'. I recall the prayer of Jonah, who is every whaler's biblical brother: 'Therefore now, O Lord, take, I beseech thee, my life from me; for it is better to die than to live.'

Will a day ever go by in which I do not chart a new course for my boy, set his compass to a new direction and imagine he is spared his meeting with a convict's murderous lust? I ask myself why the world does not open its jaws like a whale and swallow this George Nixon in one moment of annihilation!

I thought I had pursued the Evil One and looked it in the eye as I darted the killing lance into my sea-going prey ... but now I know that, all the time, the Devil was a two-legged landed creature. I have taken part in hunting so many whales; there have been so many killings at sea. Now I ask myself if it is some sort of retribution that my own son should be hunted down and pierced through the very heart himself. There are, indeed, shoals, quicksands and death-pointed rocks upon the land as well as in the ocean.

FEBRUARY 1858

For fully a year after Henry's brutal murder I could not unburden myself of a constant and enfeebling belief. I felt that the tide of my life was ebbing and that the depths beckoned. An ocean of faith that once had buoyed me up had now dwindled to a thin and muddy estuary. I was consumed by thoughts that sailed me perilously close to madness and had me believing we are merely the playthings of fate.

I believed I could see portents and omens in my life, even events from my own childhood, which presaged Henry's murder. I found myself questioning my lifetime of slaughter on the seas and imagined that now it had been repaid in bitter kind. The striving after life which once so fully marked my days stopped with an abrupt and terrible finality.

But gradually I sailed into calmer waters, turned my vision from these inner demons to the outside world and embraced the life and family that were still left to me. It was then that I came upon this *Moby Dick,* an emissary from the other side of the world which was both welcome and disturbing, a book which hinted at strange parallels in my own life.

Thus, those same Fates whom I had railed against cast me in a new drama. Throughout life they place all of us in different roles; some are shabby and insignificant, while others reveal the magnificence of high drama. Different times in our lives bring easier parts. And so it was that when I was labouring under my life's most bitter role, the heavens sent me a startling opportunity...

Late last month, at the request of our Colonial Governmment, I was engaged to man a whaleboat and, with its five crew, search out survivors from the wreck of the *Varna*. The task was not without its perils and entailed a long haul westward around a quarter of the island to somewhere near Point Hibbs. Here the British cargo ship had come to grief.

It had originally sailed from England on 12th August 1857, under the command of Captain Duncan Fleming. When he died suddenly, almost four months later, the first mate proved quite inadequate for the commanding task that was thrust upon him. On 24th December, about three miles off our west coast, the *Varna* floundered in a heavy south-west gale. She would neither stay nor wear and as she was going rapidly to leeward her grounding was inevitable. More due to luck than to skill, all hands were landed safely and they even recovered some stores before the vessel began to break up. All but four of the survivors eventually set out in the *Varna*'s longboat for Hobart-town.

The ceaseless activity, and the apprehensions I had of rescuing the remaining men let alone ensuring the safety of my own crew, lifted my eyes beyond the cruel clouds which had blighted my horizon throughout the past year. After we found the wreck we lit beacons on the shore to attract the missing quartet. In helping to save three of these men I am, in short, convinced that I too have been saved. My rescue from the twin devils of grief and remorse that have burdened my spirit throughout a year's mourning has now been effected.

In days past it was the fashion to speak of the sea as the 'primum mobile', as the source of all human health and as a fundamental element of our being. In returning to that source, I give thanks to its healing strength and attest to its beneficence.

My anguish has roared inside me like the ocean fighting at the land. But now I know that this inner sea must be stilled. I must venture outside again into the sunshine which is so much at odds with the darkness within. My family must allow their hearts to continue their endless voyage towards love. If I have learned one thing from my long and stormy existence then it is this: that we all die and are reborn several times throughout our life. The mingled threads which our days weave into this existence are calms crossed by storms until finally, at voyage's end, there lies the harbour wherein we make final mooring.

EPILOGUE

Captain William Chamberlain died in Hobart-town on 8th February 1870, aged sixty-seven. On that day, in a fitting and touching tribute to this great seaman and respected Colonist, all the ships in the port lowered their flags to half-mast.

A NOTE ON SOURCES AND REFERENCES

Remarkably, it was not until 1949 that William Chamberlain was belatedly accorded passing recognition and acclaim — in Will Lawson's *Blue Gum Clippers & Whale Ships of Tasmania*. Unfortunately, the book suffers from serious errors in its very brief biographical information: William Chamberlain was not born in 1804; his initial voyage from Sydney was not in a vessel with a letter of marque; he was educated in Scotland before, and not after, his enlistment; he did not return home 'some years later' after the Battle of Algiers.

By an extraordinary historical coincidence, when Lawson's book was being prepared one last direct link with William Chamberlain's pioneer whaling days was still clearly visible. This was the hulk of the whaleship *Marie Laure,* originally launched in 1840. At one time it was owned by William Chamberlain and, remarkably, was still afloat in Melbourne in 1948.

Two scholarly works which greatly assisted my research were *The South Sea Whaler — An Annotated Bibliography* [1985] and *More South Sea Whaling* [1991], both by Honore Forster.

PROLOGUE: THE DEVIL IN SYDNEY COVE: The whale attack in Sydney Harbour in 1790 is described in a number of first-hand accounts. Apart from these journals by First Fleeters, I have resorted to James O'Hara's *History of New South Wales* [1817].

The Aboriginal rock carving of the whale and its subsequent destruction in the mid-19th century is mentioned in Samuel Bennett's *History of Australian Discovery and Colonisation* [1867].

ASSISTANCE & PROTECTION: The Admiralty's letter to Governor Macquarie which accompanied William Chamberlain back to Port Jackson on the *Morley* and requested 'assistance & protection' is reprinted in the *Historical Records of Australia,* series 1 v. 9 [1917]. Here also will be found Dr Quarrier's letter to Earl Bathurst as well as Chamberlain's interview before the Navy Board. All originally date from late 1816.

THOUGH LOST, YET HE IS FOUND: The *Morley* did make a record-breaking run to Australia in 1816-17 when it carried young William Chamberlain home. See *The Convict Ships,* 1787-1868 by Charles Bateson [1969].

AN ACCOUNT BY DR DANIEL QUARRIER: My reconstruction of the Battle of Algiers draws on a variety of contemporary accounts as well as retrospective analyses. Foremost among these is the eminently readable, though rare, first-hand account by the Admiralty's interpreter Abraham Salame: *A Narrative of the Expedition to Algiers in the Year* 1816 ... [1819]. Also helpful was Signor Pananti's *Narrative of a residence in Algiers* [1818]. *The life of Admiral Viscount Exmouth* by Edward Osler [1835] contains a number of very significant Appendices which include documents not reproduced elsewhere.

The Battle of Algiers remains one of the least known of the Royal Navy's many great actions during the age of sail. The only recent full-length account is *Gunfire in Barbary* by Roger Perkins and Captain K.J. Douglas-Morris [1982].

Dr Quarrier did, as I noted in the text, submit his required account of the Battle of Algiers to the Admiralty. This 'Report of the state of the wounded on board His Majesty's Ship *Leander*' was published in *Medico-Chirurgical Transactions*, v. 8 [1817]. Some of his exact wording has been used in my reconstruction of the Battle and in imagining William Chamberlain's own heroic role.

Lord Minimus by Nick Page [2001] is an engaging biography of Captain Jeffrey Hudson.

A PRAYER IN BATTLE: John Campbell was indeed a devoted secretary to Governor Macquarie. His involvement as tutor to the young William Chamberlain is entirely, but not inappropriately, imaginary.

FEASTING ON THE WHALE: Apart from First Fleet journals and James O'Hara's work [already noted], the most detailed interpretation of the whale feast and its violent aftermath is offered in Keith Vincent Smith's *Bennelong* [2001].

WILLIAM CHAMBERLAIN BY JOHN CAMPBELL: See *Sydney Gazette*, 21 September 1811 for Alexander Bodie's notice re deserters from the *Frederick*. The dramatic narrative devices I have used to 'explain' William Chamberlain's 'abduction' on the *Frederick* are, of course, imaginative. No such detailed account exists. Equally, my 'explanation' for Captain Bodie's insanity is conjectural.

WILLIAM BODIE'S JOURNAL: TIMOR TO MAURITIUS: Some of the sources I have drawn on for this part of the voyage of the *Frederick* are Isaac Wellbank's *Complete Sailing Directions for the Various Passages to and through Torres 'Straits'* ... [1864] and Joseph Jukes's *Narrative of the Surveying Voyage of H.M.S. Fly* ... [1847].

The Devil of the Waters: The appalling massacre of Captain Bodie and his brother is, unfortunately, all too true. The first account to reach Sydney was published in the *Sydney Gazette* on 14 May 1814.

Strands of Rumour: The sad truth about Captain Bodie's serious derangement is [all too briefly] revealed in the unpublished papers of Governor Macquarie held in the Mitchell Library, State Library of New South Wales [Sydney], at CY 762. A remarkably similar true-life account of another insane whaling captain [Noah Folger], which I have drawn on, is given in the Old Bailey Session Papers for 11-15 April 1833.

The Whaleship Essex: The most recent account of the *Essex* disaster, a direct inspiration for *Moby Dick*, is found in *In the Heart of the Sea* by Nathaniel Philbrick [2000]. This should be read in conjunction with *Stove by a Whale* by Thomas Heffernan [1981], since it reprints a number of contemporary sources.

The little known fact that the final three survivors of this terrible ordeal were rescued and brought to Sydney in 1821 is confirmed in the unpublished log of the *Surry*. This is presumably by the ship's first mate, Edward Dobson, and is held in the Mitchell Library, State Library of New South Wales [Sydney], as Manuscript A 131.

See also *Sydney Gazette*, 9 June 1821.

A General Exodus: The annual Native meetings inaugurated by Governor Macquarie are discussed in *The Parramatta Native Institution and the Black Town* by J. Brook and J.L. Kohen [1991]. I have also drawn on the *Sydney Gazette* for 3 January 1818; 2 January 1819; 1 January 1820; 30 December 1820; 29 December 1821; 15 February 1822.

The Governor's Farewell: Two biographies with much detail of this fascinating period in colonial history were valuable: *Lachlan Macquarie* by M.H. Ellis [1958] and *Lachlan Macquarie* by John Ritchie [1986]. See also *Sydney Gazette*, 1 December 1821.

The Frederick Sails Again: There was indeed a convict seizure of the brig *Frederick* [not Captain Bodie's vessel] from Macquarie Harbour, Tasmania, in William Chamberlain's day. I have taken the liberty of slightly altering dates here to suit the purposes of narrative drama. The actual event occurred early in January 1834, with its desperate pirates being shipwrecked as far away as South America in late February. See *The Capture of the Frederick* by James Porter [1981].

Alexander Pearce's descent into cannibalism after his escape from Macquarie Harbour is told in *Hell's Gates* by Paul Collins [2002].

Scrimshaw: The actual production of the 'South Polar Expedition' in Hobart opened a year earlier than I have noted, on 3 May 1841. The expedition had arrived in Hobart-town on 17 August 1840, stayed until early November and returned in late April 1841.

See two articles by Gillian Winter: 'Tasmania hails you as her favour'd guest', *Tasmanian Historical Research Association Papers and Proceedings*, v. 38, nos. 3 & 4 [December 1991], and 'South Polar Expedition' in *Companion to Theatre in Australia*, ed. Philip Parsons [1995].

Days Ashore: The description of the theatrical production 'The Devil & the Deep Blue Sea' is, while hopefully entertaining, nevertheless imaginary.

The Bun-yip: Elsewhere I have written at length about the exhibition of the supposed 'bun-yip' skull in Colonial Sydney in 1847. See my *Bunyips: Australia's Folklore of Fear* [2001]. While Alexander

Macleay certainly refuted the skull's exotic claims, his meeting with William Chamberlain is imaginary. Surprisingly, there has been no full-length biography of the Macleay family to date. Among various articles, one by Julian Holland is of particular interest: 'Diminishing circles: W.S. Macleay in Sydney, 1839-1865', in *Historical Records of Australian Science*, v. 11 no. 2 [1996].

William Chamberlain's quotation from Sir William Jardine's writing comes from Volume 6 of the 'Mammalia' series of his popular *Naturalists' Library*. This volume was published in 1837 and was devoted to whales.

THE WONDERS OF EDEN: William Chamberlain refers to an item on Captain Raine in the *Sydney Gazette*; it occurs on 15 August 1828.

For the remarkable yet true account of whaling at Twofold Bay I have drawn on the first-hand narrative by Oswald Brierly in 'Reminiscences of the sea', Mitchell Library, State Library of New South Wales, Manuscript A 546. Also, 'Whaling at Eden', by H.S. Hawkins and R.H. Cook in *The Lone Hand*, v. 3 no. 15 [July 1908].

INSIDE THE WHALE: A whale's skeleton was, surprisingly, exhibited outside the Australian Museum in 1850. Its extraordinary story is outlined in an anonymous pamphlet usually attributed to William Wall but almost certainly by Alexander Macleay: *History and Description of the Skeleton of a New Sperm Whale* [1851].

THE DEVIL ON LAND: Regrettably, all the quotations relating to Henry Chamberlain's murder are true and are transcribed from the original Colonial newspapers of 1856 and 1857.

VOYAGE'S END: Details on the wreck of the *Varna* are given in the *Hobart Mercury*, 23 January 1858; the *Hobart Town Courier*, 25 January 1858, 10 February 1858 and 12 February 1858; the *Argus* [Melbourne], 1 February 1858. See also *Tasmanian Shipwrecks, v.* 1,

1797-1899, by Graeme Broxham and Michael Nash [1998]; *Australian Shipwrecks, v. 2, 1851 to 1871*, by Jack Loney [1980].

Captain Chamberlain commanded the *Maria Orr,* which was launched on 16 January 1838 as the first full-rigged whaleship to be built in Hobart. In recognition of this distinction the Tasmanian Government presented her with a suit of sails. I have imagined the grieving father shrouding his son's body in a remnant of this treasured cloth.

EPILOGUE: The funeral tribute to William Chamberlain is taken from the [Hobart] *Mercury,* 9 February 1870.

ACKNOWLEDGEMENTS

Any writer, like any sailor, knows that it is near impossible to write a book or sail a ship alone. One needs a crew — volunteer, enlisted, impressed or otherwise — in order to stay afloat and bring the voyage safely to an end. For my part I was fortunate in being able to assemble a seasoned crew who helped bring this work through to its final mooring.

Some of the incidents and adventures used to flesh out the adult life of William Chamberlain have been culled from a variety of first-hand accounts of nineteenth-century whaling. While hopefully they give an air of verisimilitude to these years, it is worth observing that none could really compete with the extraordinary and elusively documented events that occurred in his younger days.

I am indebted to many friends and colleagues in libraries throughout Australia for their assistance in locating these texts.

At the beginning I was given very significant impetus by Dr Rory Muir, whose bibliographical help with the Battle of Algiers was of inestimable value. Cassie Cassingham and the staff of the State Library of Queensland [Brisbane] harpooned valuable tomes for me from their Eber Bunker Maritime Collection. Richard Overell, Rare

Books Librarian, Monash University Library [Melbourne], kindly supplied a copy of Dr Daniel Quarrier's 'Report' from *Medico-Chirurgical Transactions* [1817]. Joan Keating, Special Collections Librarian, University of Queensland Library [Brisbane], extended a warm welcome to an old colleague and shared the rare copy of Abraham Salame's *Narrative of the Expedition to Algiers* [1819] from the Father Leo Hayes Collection. The staff and the archivist at the Australian Museum Library [Sydney] were unfailingly helpful to a grateful colleague. Dr Ronald Strahan [formerly of the Australian Museum] confirmed my suspicion about Alexander Macleay's authorship of the 1851 pamphlet on the skeleton of the whale. Des Cowley, Rare Books Librarian, State Library of Victoria [Melbourne], and Warwick Hirst and Stephen Martin, State Library of New South Wales [Sydney], were helpful colleagues in a time of need. The staff of the National Maritime Museum Library [Sydney] and of the Duntroon Military Academy Library [Canberra] are also gratefully acknowledged. The redoubtable Aileen Brown and Anne Jackson introduced me to Marten Syme of Port Fairy. I am also indebted to Susan Geason for professional advice and especially for empathy from a much respected fellow writer. Diana Chicchio remained a staunch and enthusiastic friend throughout the writing of this book.

And Fiona Inglis, my literary agent at Curtis Brown Australia, helped land this big fish with HarperCollins!

ABOUT THE AUTHOR

Robert Holden was born in Brisbane in 1948, an identical twin. He now lives in Sydney. He is married, with a 24-year-old son, Nicholas, who was recently the joint author with his father of *Bunyips: Australia's Folklore of Fear*. Robert Holden is well known as a lecturer and guest speaker in Australia, New Zealand and England. His wide-ranging academic qualifications have enabled him to curate successful and high-profile exhibitions across various disciplines, including Australian fine arts, bibliographical history and Australian cultural and social history. These exhibitions have shown in national, state and regional venues throughout Australia as well as in the Victoria and Albert Museum, London and the Australian Embassy, Paris.

Robert Holden is the author of twenty books to date, most recently *Orphans of History: The Forgotten Children of the First Fleet* and *Bunyips: Australia's Folklore of Fear*. The latter was produced in association with the National Library of Australia and accompanied by a nationally touring exhibition. *Orphans of History* marked a change in direction for Holden, away from art and cultural studies into more mainstream history. This book was awarded a senior fellowship from the Literature Board of the Australia Council and film rights have been optioned by Paramount Studios with a screenplay by David Williamson.